CW01658854

CONTESTING PUBLIC SPACES

This book explores concerns for spatial justice as streets, squares, and neighbourhoods are continuously made and remade through planning processes, political ambitions and everyday activities. By investigating three sites in London that have been the focus of masterplanning, Ed Wall exposes conflicts between planning offices and private developers who direct large urban change and community groups, market traders and residents whose public lives are inseparable from their neighbourhoods being reconfigured.

The book uniquely brings sociological approaches to what are often considered architectural concerns, revealing challenges as London's public spaces are designed, regulated and lived. Through in-depth research, Ed Wall identifies how uncertainty caused by large-scale urban strategies, the realisation of visual priorities, and uneven relations between private interests, public organisations and daily lives determine the public realm of global cities.

This work is intended for readers interested in how the urban spaces of their cities are continually produced in competing ways—from architecture and urban studies scholars to planners and politicians.

Ed Wall is Academic Lead of Landscape Architecture and Urbanism and co-director of the Advanced Urban (AU) research group at the University of Greenwich, London. He is a visiting professor at Politecnico di Milano, and in 2017, was City of Vienna Visiting Professor: Urban culture, public space and the future—urban equity and the global agenda (SKuOR/TU Wien). His research and design work explores practices of public space and processes of landscapes through concerns for spatial justice.

CONTESTING PUBLIC SPACES

Social Lives of Urban Redevelopment in London

Ed Wall

Routledge
Taylor & Francis Group
NEW YORK AND LONDON

Cover image: Ed Wall

First published 2022
by Routledge
605 Third Avenue, New York, NY 10158

and by Routledge
4 Park Square, Milton Park, Abingdon, Oxon, OX14 4RN

Routledge is an imprint of the Taylor & Francis Group, an informa business

Library of Congress Cataloging-in-Publication Data
A catalog record for this title has been requested

ISBN: 9781032163574 (hbk)
ISBN: 9781032163567 (pbk)
ISBN: 9781003248187 (ebk)

DOI: 10.4324/9781003248187

Typeset in Bembo
by Apex CoVantage, LLC

To my parents, Rosie and Tony, and my grandparents, for sharing their interests in learning

CONTENTS

FIGURES

PREFACE

When I began exploring the contestation of London's public spaces as they were planned, managed and lived, I did not expect accounts from conversations and interviews in planning offices, architecture studios, neighbourhood cafes and open spaces to resonate so loudly. In South London's Elephant and Castle, concerns of the market manager who did not know if the low-cost market would be demolished to make way for a new urban development contrasted with the knowledge of developers that their masterplan would unsettle many lives and destroy long-established businesses. In Paddington, accounts of homeless people displaced from public spaces and local hostels were accepted by a local government officer as 'social cleansing' and rationalised as a function of how the market worked. In Central London, conflicting claims of leading architectural practices to design landmark public spaces highlighted the potential gains of being associated with the remaking of London's globally recognisable civic squares. These accounts of transforming public spaces, shared by residents, market traders, developers, planners and politicians, reflect a comment offered by an historian who had witnessed his neighbourhood comprehensibly reconfigured twice through large-scale masterplanning: 'Any major redevelopment naturally produces a fair amount of controversy, especially if the world detects any unfairness.'

As I bring together narratives of public spaces, all that have been the focus of large-scale masterplanning, I reveal concerns for fairness in how these places are made and remade. By examining relations between contrasting accounts of making public spaces—approaches to public space that frequently come into contact, reconfigure and conflict—I have identified what can be gained and lost as London's public realm is redesigned, maintained and lived. I examine concerns of local market traders and the operations of business owners as they are undermined by masterplans set out by local government planners. I describe approaches of internationally renowned architects fulfilling the briefs of commercial developers and corporate clients. I highlight globally recognisable events planned for London's central civic spaces and the significance of producing imageable landscapes through practices of architectural design. By bringing these accounts together I reveal tensions between many ways that the public realm is produced—from planned to lived, from owned to managed, from programmed to policed.

It is the personal descriptions of roles and understanding of how public spaces in London are made that have formed this book, so I would like to express my thanks to the residents, tourists, cleaners, security guards, planners, architects, urban designers, historians, market-stall holders, landlords, public space managers, developers and politicians who shared their time. The deepest thanks to the low-wage workers and market-stall holders whose lack of influence within the masterplanning processes never dented their willingness to talk with me. I am also grateful to the openness of the individuals who described personal struggles within competing agendas for London's public spaces.

I am indebted to Fran Tonkiss and Suzi Hall at the London School of Economics and Political Science (LSE) Cities Programme, whose careful readings and supportive critiques have been invaluable throughout this research. Their patience and direction as I have navigated social science methods and their encouragement for me to employ my architectural knowledge provided an essential framework from which my research developed. Thank you also to Ben Campkin, Monica Degan, David Madden and Ricky Burdett for their advice as I progressed through MPhil and PhD examinations. I appreciate the support from the wider community at the LSE, from the award of a London School of Economics PhD Scholarship to exchanges with colleagues in Sociology, the Cities Programme and the NYLON network. I benefited greatly from working alongside and following colleagues, especially Katherine Robinson, Torsten Schroeder, Günter Gassner and Adam Kaasa. I would especially like to thank Francisco Calafate-Faria, for revealing areas of overlap between my research and my earlier work in landscape architecture and urban design, Antonia Dawes, for generous discussions about market stalls and ethnography, and Michaela Muscat for advice on social science theory.

I am grateful for the support and inspiration of all my academic colleagues, students, family and friends, of which there are too many to list here. However, several individuals must be thanked: The inspirational writing, unequivocally positive references and endless encouragement that Michael Sorkin and Grahame Shane offered have formed the foundation of my academic world. I would like to thank Colin Marx for his reading of my research proposal and Ed Soja for his generous advice on early investigations. Antonella Contin and colleagues at Politecnico di Milano have consistently afforded me opportunities to discuss this research through teaching and research over the last ten years while colleagues at the University of Greenwich, notably Nic Clear and Neil Spiller, encouraged me to incorporate my research into programmes and teaching. Tim Waterman and Jane Wolff generously commented on some of the Elephant and Castle research in 'Working with uncertainties', included in our edited collection *Landscape Citizenships* (Waterman et al. 2021). The support of Sabine Knierbein and colleagues from SKuOR at TU Wien has been invaluable in sharpening my critique of public space and expanding my knowledge in urban studies. Additionally, all the students in London, Milan and Vienna with whom I have worked have provided an essential testing ground for the development of my conceptions of publicness. In particular, I would like to thank the students involved in development of The Department of Public Space, an experimental recoding of public space that along with Michael Sorkin's *Local Code* (1993), informed the concluding epilogue.

I continue to be thankful for the commitment of Routledge to publishing my writing, especially Kathryn Schell, Sean Speers, Christine Bondira, and Megha Patel who have supported the development of this book.

Finally, I would like to thank Kristin for encouraging me to embrace the opportunity to undertake this research and Carly whose endless joy pushes me to continue.

INTRODUCTION

Contesting public spaces

London's public spaces embody a constant reworking of relations between economic agendas, government departments, public agencies, competing businesses, daily routines and its urban fabric, frequently mediated through masterplans for redevelopment. The city's public spaces reflect Lyn Staeheli and Don Mitchell's description of 'the public nature of space' that is never settled but is 'continually made and remade' (2008:xxii). Interwoven and competing ambitions, from urban designers setting out proposals to transform London's architectural spaces to individuals reconstituting the city through their daily activities are unevenly realised as urban developments unfold. Since the late 1990s new urban development policies and intensified economic agendas have led local councils and the Greater London Authority (GLA), the metropolitan government that was formed in 2000, to facilitate many neighbourhood-scale transformations across the capital. However, the masterplans, which are largely informed by private interests and international investors and which produce highly managed and carefully imaged public spaces, are frequently challenged by individuals and community groups.

In *Contesting Public Spaces* I explore conflicting accounts of how three public spaces in London have been made and remade, each within large-scale masterplanning projects. I examine relations between contrasting ways that public spaces are produced and I question what is at stake as unequal opportunities, from political strategies to small-scale and unplanned actions, are taken and contested. I investigate the impact on individuals whose lives are part of the public spaces of these masterplanned areas as well as the priorities of organisations whose work is involved in their transformation. I argue that claims by politicians, developers and architects to make new public spaces as architectural forms, along with desires to share tourist photos and redevelopment images, are frequently prioritised over public spaces that accommodate a diversity of everyday activities. *Contesting Public*

DOI: 10.4324/9781003248187-1

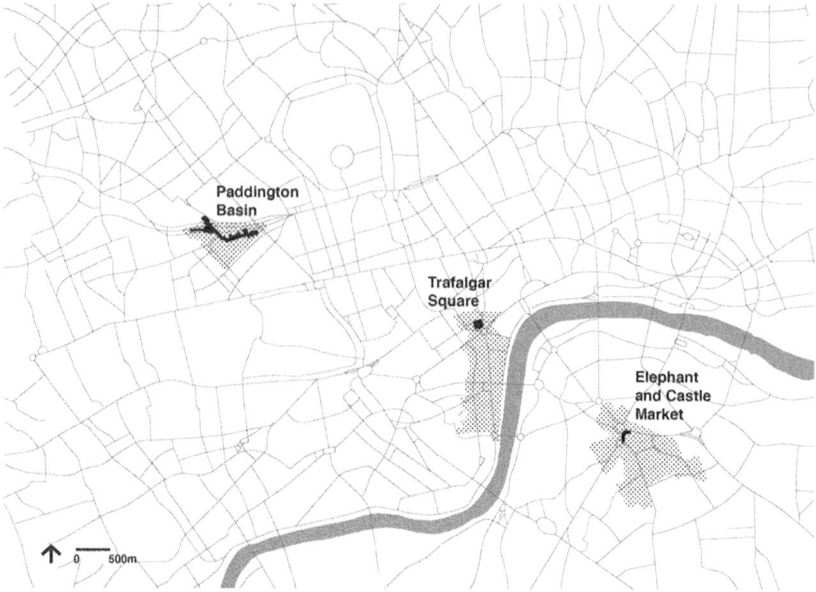

FIGURE 0.1 Three London sites of public space within the context of large-scale masterplanning.

(Map by author)

Spaces highlights how people and uses are excluded and smaller business operations are closed down as unequal relations that come to produce new public spaces serve the economic ambitions of developers, public space managers and governments.

In this chapter, I introduce the three London public spaces within the contexts of large masterplanning sites and long timelines of planning and construction. I outline the intersection of processes of urban redevelopment, practices of public space and contrasting London contexts, each that are unpacked in greater detail in Chapters 2, 3 and 4 respectively. I also explain the research approach that I have employed, bringing together architectural and social science perspectives on public space, and the questions I asked that reveal London public spaces contested as they are made and remade.

Three public spaces

In this book, I focus on three London sites[1] that have been subject to such large-scale change: the deteriorated sunken plaza of Elephant and Castle Market that wrapped around the now demolished shopping centre; the highly managed corporate canal-side of Paddington Basin that was constructed across a former industrial area; and the redesigned grand civic space of Trafalgar Square in the centre of London, a place that attracts thousands of visitors each day. Each landscape has been the focus of masterplanned redevelopment over the last three decades: *Elephant and Castle Regeneration* (2004–2029), *Paddington Waterside* (1998–2018) and *World Squares for All* (1996–2003). The large-scale developments have restructured local economies and attracted the attention of international investors. Their implementation has led to transformed skylines with new buildings, reconfigured roadways, demolished neighbourhoods, displaced businesses and relocated residents. The masterplans have also redefined relationships, physical forms and activities of public spaces through contrasting planning, economic and design ambitions that are in play across London: at Elephant and Castle, an unbalanced partnership between the local council and global property developers to comprehensively demolish and reconstruct the area; at Paddington Basin, a development led by commercial developers and facilitated by the local planning authority to transform a brownfield site, creating highly controlled public spaces; and at Trafalgar Square, a central government initiative embraced by the GLA to reinvigorate a central London landmark and redefine the image of London through programmed events.

The public spaces of Elephant and Castle Market, Paddington Basin and Trafalgar Square are explored as publicly accessible sites designed, managed and used by a variety of citizens, communities and organisations. They are places appropriated through redevelopment and claimed as public spaces during planning processes. They are also public spaces that in varying ways are taken back and made public through occupation, use and adaptation. Throughout this book I adopt multiple, often conflicting, definitions of public spaces as physical spaces, often formed by developers, consultants and government departments, as well as social spaces that are constantly reconfigured by individuals who use these sites for routine activities and to gather around public concerns. These are places defined by teenagers meeting up after school; vendors setting up small businesses; students gathering around educational concerns to protest in London's streets; and pension companies using their financial means to profit from new public squares. I investigate how these public spaces are formed through financial agreements in planning negotiations, design decisions by architects and the daily routines of each site. These are contrasting claims to public spaces, produced from relations between government departments, local authorities, property developers, global investors, cultural commentators, Business Improvement Districts (BIDs), the Metropolitan Police, public space managers, architectural consultants, activists, residents and visitors—revealing contestations between how public spaces are developed, managed and used.

I recognise public spaces, as Doreen Massey describes, 'as the product of social relations which are most likely conflicting and unequal' (2005:152). For organisations, groups and individuals involved in redeveloping, operating and occupying these public spaces there are conflicting issues of reputation, profit, access and use at stake. The contrasting approaches to making public spaces—from practices of top-down redevelopment that Michel de Certeau describes as a 'projection that is a way of keeping aloof' (1984:92) to close-up interactions in the street 'below thresholds at which visibility begins' (ibid.:93)—are core concerns in this book.

At Elephant and Castle Market the ambitions of planners who wished to reimagine the area conflicted with commercial interests of the adjacent shopping centre owners. Years of uncertainty over the future of the regeneration of Elephant and Castle left market stall holders operating in increasingly deteriorated conditions, as the market managers and the stall holders struggled to plan the future of their businesses. As the old low-cost market was planned to be demolished and a new square was designed to host a craft market, doubts were raised as to whether the existing market traders should be allowed to continue in the new space. At Paddington Basin the primary objective of the Canal and River Trust, the charity that owns the canal basin for 'public benefit' (www.canalrivertrust.org.uk 2014), contradicted the operations of the commercial public space managers who oversaw the towpaths on behalf of private leaseholders. Narrow terms of publicness, which included highly restricted uses and occasional evictions of those entering the space, were compounded by increased policing across the Paddington area and expanded influence of the development owners through the establishment of BIDs. At Trafalgar Square many groups sought a public space open for debate and representation in which to gather around shared concerns such as global warming, social justice and fair education. At the same time the GLA that manages the square had imperatives for visual aesthetics, health, prosperity, connectivity, security and world city status (Great London Authority (GLA) 2009): the first of the GLA's objectives in their *Manifesto for Public Space* was to create a 'beautiful city . . . with spaces that are fit for a world city' (GLA 2009). The use of the square for public actions remains only part of a schedule of political, cultural and commercial events that exclude less desirable activities of busking, vending and sleeping.

Redevelopment of public space

Elephant and Castle Market

Three cases of redevelopment frame the accounts of making public spaces. Elephant and Castle Market is at the centre of the *Elephant and Castle Regeneration* masterplan, a transformation that aimed to reconfigure Elephant and Castle through a partnering of the local authority with a private developer. Involving private interests in urban redevelopment was promoted by the Conservative government from the 1980s and then advanced by New Labour when they were elected in 1997. At Elephant and Castle this focused on a £1.5bn regeneration around a strategic node in South London. The regeneration was based on an agreement between the London Borough of Southwark, as a public authority, and Lend Lease Corporation Limited, global developers. The plan proposed to develop the GLA designated Elephant and Castle Opportunity Area (2004), fifty-five acres around Elephant and Castle, including: the Heygate Estate; transport interchanges; and the Elephant and Castle Shopping Centre around which wrapped the market. Since 1990 the shopping centre and the market had been

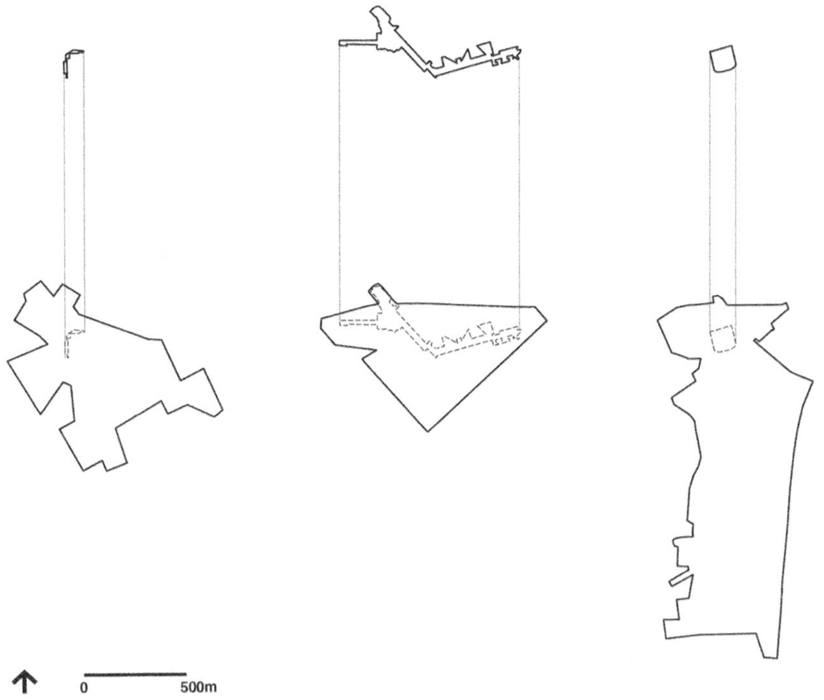

FIGURE 0.2 Elephant and Castle Market, Paddington Basin and Trafalgar Square in the context of *Elephant and Castle Regeneration*, *Paddington Waterside*, and *World Squares for All* masterplans.

(Diagram by author)

through repeated attempts at regeneration by its owners. Then following the financial crisis of 2008 many of the planning agreements were thrown into disarray. By 2013, the shopping centre and the market was again the focus of discussions about demolition, with Councillor Fiona Colley, cabinet member for regeneration at Southwark, describing the site as 'the last piece in the jigsaw for the regeneration' of Elephant and Castle (www.southwark.gov.uk 2013).

Before the regeneration began, the Elephant and Castle Market was a place animated by games of checkers in the food court, market stalls constructed and dismantled each day and customary conversations as low-cost produce was sold. Along with the shopping centre, the market was the centre of Elephant and Castle's public life. Although neither the shopping centre nor the market were owned by the development partners, Southwark Council and Lend Lease, they were included in the 2012 Supplementary Planning Document that set out the plan for demolition and replanning. The removal of the shopping centre was essential to the council's ambitions at Elephant and Castle. Ben Campkin recognises in *Remaking London* that 'more recent public debates . . . have centred most prominently on the shopping centre' (2013:68). However, less focus was afforded to the market that was tied, through common landownership and short-term leases, to the plans for the shopping centre. The overlooked, low-cost and privately operated market was a hub of commercial and public activity at Elephant and Castle for over three decades. Over the final ten years, before it was demolished in 2021, uneven relationships between organisations leading the regeneration and individuals impacted were increasingly evident.

The redevelopment of Elephant and Castle existed within a *Local Authority London*, highlighting the need of local councils to cede land and responsibility for development through partnering with private interests. Successive political agendas have, as Susan Fainstein describes, 'promoted physical change with the expectations that better-looking cities are also better cities' (1994:2). She continues:

> The quandary for local political officials is that they must depend on the private sector to finance most economic expansion, and they only have limited tools for attracting expansion to their jurisdictions.

By restricting the planning powers and budgets of local authorities, central governments forced councils, such as Southwark, to partner with private developers who recognised that large areas of land used for social housing, local businesses and green spaces can be appropriated for redevelopment. With Southwark Council's ambition to update its marginal neighbourhoods and deteriorated public infrastructures, but with limited options available, the form, pace and objectives of developments have increasingly reflected developer timelines, investor ambitions and global markets. Fainstein (1994:123) identifies common characteristics of such projects:

> they involve the re-use of commercial areas by different enterprises that would cater to new kinds of customers; they require a major restructuring of land uses and the relocation or closing of existing business premises; and they [are] likely to cause changes in the composition of the working and residential populations in the project area and their surroundings.

This is a London of insecure and asymmetrical partnerships, between Lend Lease Corporation, Southwark Council and other landowners, who have repeatedly renegotiated the terms of their agreements to ensure adequate profit for private developers while denying sufficient public transparency.

While developers gain through commercial leases and sales of apartments, Southwark Council receives benefits as developers provide new parks, squares and renovated Tube infrastructure. Section 106 contributions, also known as planning obligations, are negotiated between developers and local authorities to provide resources to offset the impact of developments (Town and Country Planning Act 1990). The Community Infrastructure Levy (CIL) is a more recent tariff-based approach intended to support the development of local areas. Both mechanisms are used by local authorities to provide new public spaces for the benefit of existing

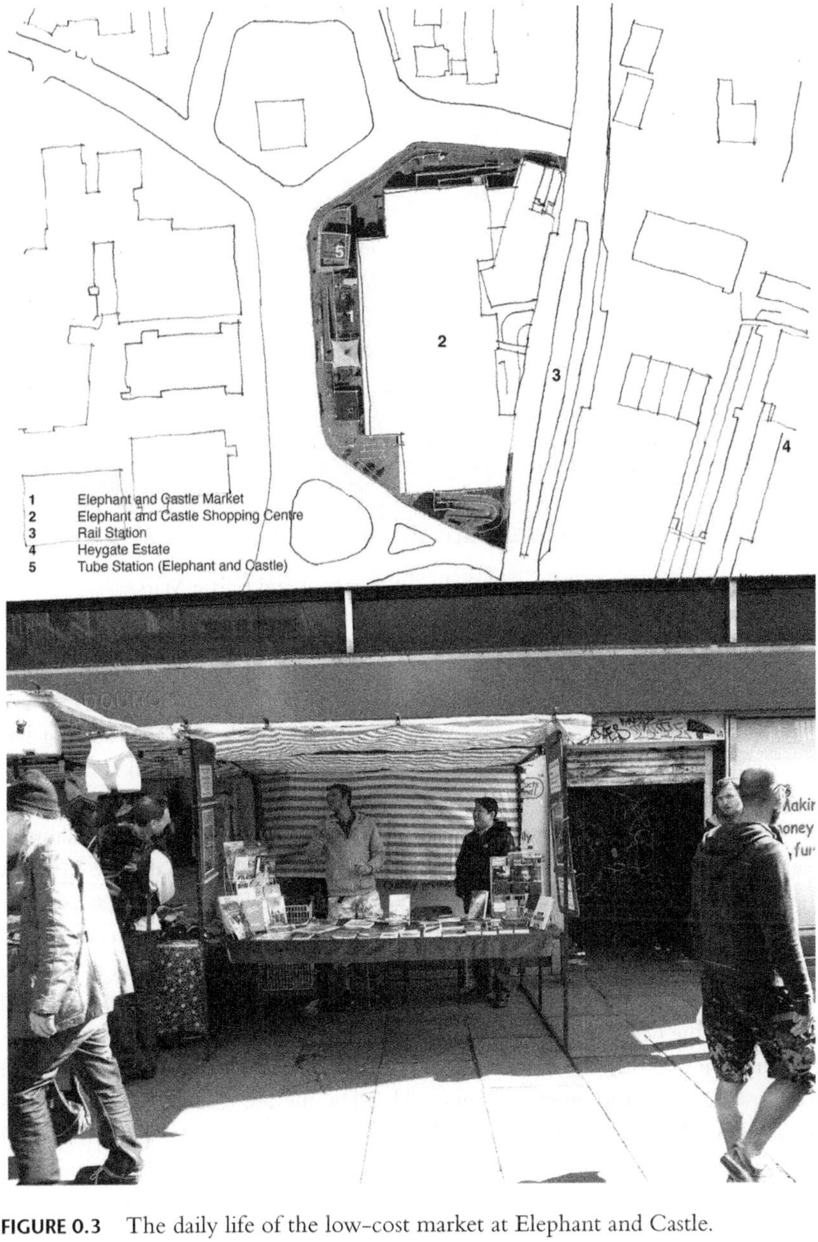

FIGURE 0.3 The daily life of the low-cost market at Elephant and Castle.

(Collage by author)

FIGURE 0.4 Traders sustaining public life of Elephant and Castle Market.

(Photo by author)

areas and communities. Southwark Council's approach has been to use Section 106 contributions 'for defined site specific mitigation' and CIL in order to 'secure contributions towards strategic infrastructure' (www.southwark.gov.uk 2014). Both planning instruments can return benefits to developers who frequently maintain control of the new public spaces, and to the local authority who are able to divest their responsibilities for ongoing costs and maintenance. But Section 106 negotiations, as those between Southwark Council and Lend Lease, are seldom conducted in public, leading to criticism by government agencies and by Elephant and Castle residents. Paul, a local resident who worked at the council averred: 'My personal view is that the people who were negotiating at a high level were not . . . transparent' (Interview 2012). The lack of information about the process of regeneration has caused suspicion amongst residents and local businesses: when private developers such as Lend Lease and Delancey take over the provision of public space from local government, they haggle over the scope and scale of their projects while appropriating control of surrounding spaces to visually foreground their developments.

Paddington Basin

The *Paddington Waterside* masterplan is the basis for fashioning open spaces around Paddington Basin, a network of canal-side spaces that has been slowly built through a process of developer-led redevelopment. The Paddington area of West London, located behind the grand rail terminal building designed by Isambard Kingdom Brunel, was designated by the City of Westminster as Paddington Special Policy Area in 1988 and was subsequently selected as an Opportunity Area by the GLA in 2004. The masterplan area includes fourteen redevelopment sites across eighty acres with open spaces of Paddington Basin at the core. This development, led by Paddington Partnership (formerly Paddington Waterside Partnership), was planned to be completed around 2018. The partnership's intention for the open spaces was to offer visitors a seamless transition from the public streets beyond Paddington Basin to the privately controlled open spaces within. But the limited number of entry points, some of which are closed at night, leave the development area less welcoming, while the activities within the canal-side spaces, the patrol of private security guards and the changes in paving and furniture reveal contrasting landscapes inside and outside the masterplan area.

Paddington Basin has been created within and by a *Corporate London*, a city of highly facilitated urban developments producing tightly regulated public spaces of streets, squares and amphitheatres that fit neatly between new offices and tall residential buildings. It is a London that has benefited from three decades of government policies, informed by economic liberalism and free markets, that can be read through the building, reconstruction and occasional abandonment of the city's public spaces. Firstly, in the 1980s, swathes of formerly industrial sites, waterfronts and rail yards, such as Paddington Basin, that had facilitated the mercantile and industrial growth of London and England, were sold for development by the government and public agencies. As the City of Westminster followed the UK government's laissez-faire approach to planning in the late 1990s, Paddington Basin was divided into development parcels which were bought, traded, subdivided and leased depending on the opportunities for profit from their fluctuating property values. Westminster initiated the masterplan and approved each proposal for planning permission. Although the development phases at Paddington Basin have progressed sporadically, there has been a consistent building-out of an infrastructure of pristine architectural open spaces. These landscapes elegantly foreground the building developments while realising ambitions for public access by the Canal and River Trust, who sold the land along the canal to developers on long leases. The featureless landscape, paved in stone with stainless steel details, is highly maintained through routines of repairing and cleaning. Claims in planning documents to the publicness of these spaces are undermined by evidence of private security guards evicting people deemed undesirable, a tactic enabled by the City of Westminster.

What are variably described as public-private, semi-public, semi-private and privatised public spaces, characterise many of the new open spaces of this *Corporate London*. The private terms of development, maintenance and security that are reinforced at Paddington Basin have created a landscape where the cost to maintain the development's new open spaces to the level desired by the developer is higher than what the local authority could provide. The development parcels are transformed

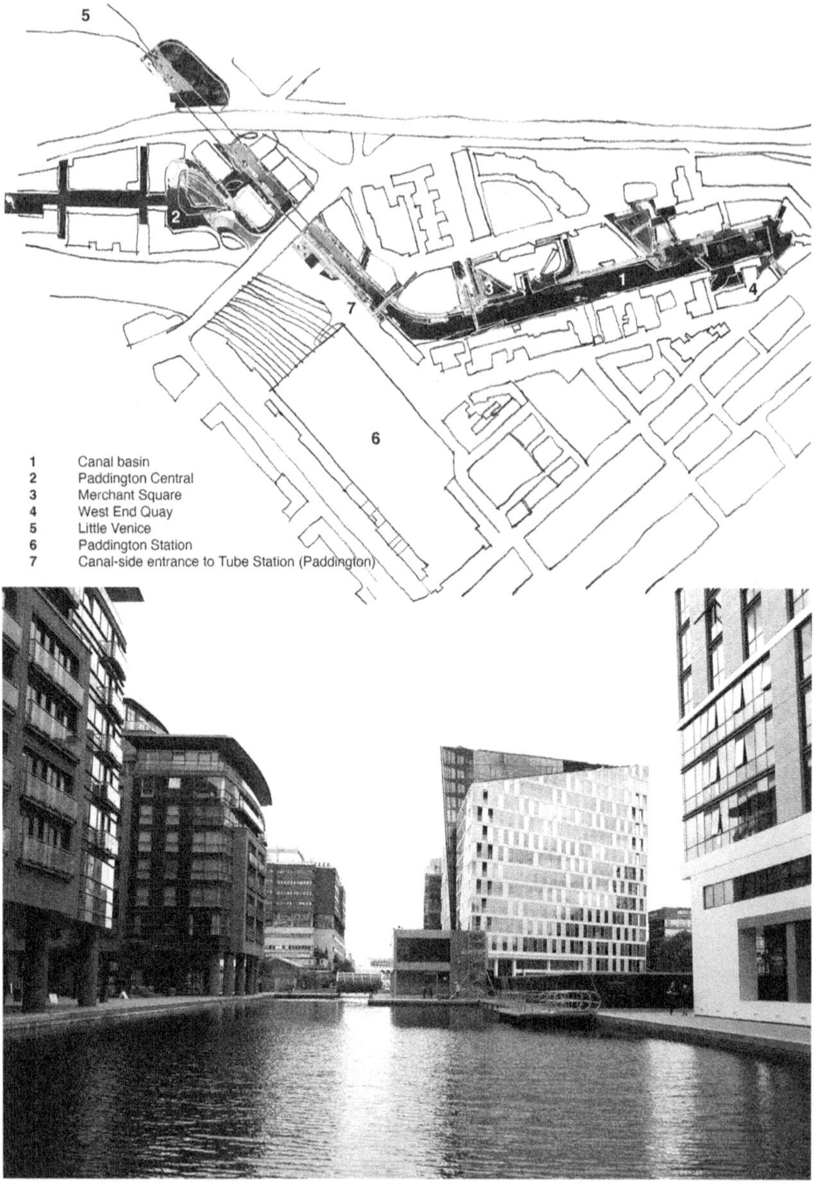

FIGURE 0.5 Unwelcoming canal-side spaces along Paddington Basin.

(Collage by author)

FIGURE 0.6 Intensely maintained and controlled spaces along Paddington Basin canal.

(Photo by author)

by private developers and, once completed, remain owned by banks, investment funds and pension holdings. They have restricted access, discourage large groups of people and they completely deny unlicensed gatherings and protests. Small stainless-steel plaques are ubiquitous across these sites, proclaiming that the spaces described as public in planning meetings are actually private and that access and use can subsequently be withheld. At Paddington Basin there are also multiple forms of management, including the presence of property management company Broadgate Estates who were early pioneers in the management of privately owned public spaces in the UK in the 1980s. Privately dominated changes to how public open spaces of this *Corporate London* are managed continue unmitigated despite a study commissioned by the former Deputy Mayor, Nicky Gavron, stating:

> London planning has become adept at the delivery of high quality public realm as part of large scale private developments . . . the planning system has not caught up with the fact that what is in effect public space is often subject to private management.
>
> *(GLA 2011:9)*

Corporate London has also embraced BIDs since they were introduced in the UK in 2003. BID organisations have been established to manage neighbourhoods across London addressing inconsistencies and frequent reductions in government funding of public spaces since the 1980s. The deployment of BIDs was questioned in early discussions around the privatisation of public space (Sorkin 1992) and continue to be criticised in North America as focusing on 'consumption, gentrification, and the maintenance of a "pro-business" streetscape' (Madden 2010:196). The main differences between BIDs in the UK and the USA is that to fund the former, a levy is charged to businesses whereas in the latter the building owner is charged the BID levy. The activities of BIDs extend commercial approaches across the public realm of neighbourhoods limiting the publicness of streets, squares and canal-side spaces.

Trafalgar Square

Trafalgar Square is a key component of the *World Squares for All* masterplan that was initially commissioned by the Government Office for London in 1996. The masterplan, which includes the redevelopment of Trafalgar Square as its first phase of implementation, brought together several public organisations, such as Transport for London and the City of Westminster. The plan proposed a remodelling of Trafalgar Square and extending the civic square up to the National Gallery, achieved by the partial road closure of Pall Mall East. New steps were installed to further align the square with the gallery, while a café and public toilets were built under the upper terrace. Aside from these physical changes, the main features of this civic square remained; the two grand fountains, the four lions and Nelson's column were renovated to maintain a space and image of London and the UK.

During the period of redevelopment the management of Trafalgar Square was transferred from the City of Westminster to the GLA. Although the square remains under ownership of the Crown Estate, responsibility for the square was handed to the GLA through the Greater London Authority Act 1999. During Trafalgar Square's redevelopment several individuals were evicted through the enforcement of new regulations. Pigeons and commercial vendors were removed from the central square and a requirement for written permission to engage in public activities, such as assembly and public speech—activities with which Trafalgar Square is commonly associated—were enforced. The administration of the bylaws and the employment of Heritage Wardens, cleaners and the falconer have since created an alternative presence of contractors who patrol the space, keep it clean and ensure that it is pigeon-free. Ambitions of the GLA to attract global attention to London through its programme of events and visitor experiences highlight the role of cultural, commercial and artistic operations in remaking the square. However, this has led to concerns that the square is dominated by scheduled events, gatherings and promotions and is less available for more spontaneous activities and daily visitors.

Trafalgar Square contributes to a *Global London*, understood in terms of the events, images and aspirations for a world-class capital city promoted under the auspices of the Mayor of London and the GLA. It is a London understood

FIGURE 0.7 Programmed cultural events restrict access to London's Trafalgar Square.

(Collage by author)

FIGURE 0.8 The most photographed view of Trafalgar Square frames the fountains in the foreground and the National Gallery and St Martin-in-the-Fields church in the background.

(Photo by author)

through public spaces that have come to represent what Ash Amin describes as 'expressions of achievement and aspiration by urban leaders and visionaries' (2008:5). After the New Labour government was elected they established the GLA, which was afforded, through the 1999 Greater London Authority Act, oversight of both transportation and planning across the capital. The Mayor of London became responsible for Trafalgar Square and Parliament Square, which at the time were midway through being reimagined by the *World Squares for All* masterplan. Aspa Gospodini suggests that during this period the 'quality of urban space' became a prerequisite for attracting investment into cities like London, Paris and Berlin (2002:12). To encourage property speculation in commercial and residential areas, *Global London* is drawn into competition with other cities wishing to maintain a global status. Central government, the GLA and the office of the Mayor of London aimed to create distinct and attractive conditions for investment—frequently through London's public spaces. As former Deputy Prime Minister John Prescott stated: 'London is built around its squares and its architecture and its old buildings—and they are beautiful!' (Interview 2013).

At the heart of the *World Squares for All* masterplan are Trafalgar Square and Parliament Square, which offer a scale of uninterrupted spaces appropriate to foreground the architectural landmarks of the National Gallery and the Houses of Parliament.

In addition to reorganising Trafalgar Square, whose redesign reflects the pedestrian-oriented priorities of world-famous public space advocates and urban design consultants Jan Gehl and Richard Rogers, the square's management and operations were also rewritten by the GLA. Furthermore, the square was used to host a programme of events that further contributed to the image and experience of London. Trafalgar Square has since accommodated many media-friendly events, such as the countdown to London's 2012 Olympic Games and the shooting of the James Bond film, *Skyfall* (2012). *Global London* attracts a swelling number of tourists, competes as a cultural destination and aspires to be seen as a centre of democracy. It is situated in the centre of London, an area that attracts national and global attention and was described by the GLA in its *Manifesto for Public Space* as the 'economic engine of our city and its showcase for visitors' (GLA 2009:12). These priorities exist within a particular nexus of the city's history, where London's 'World Class' status aligned with a mayoral focus on actively making new public spaces 'fit for a world city' (GLA 2009:02).

Approaches to public space

Through this book I explore the socio-spatial dimensions of public spaces. My experience as a landscape architect and urban designer has previously led me to work with diverse spatial practitioners (such as artists, architects, engineers) and organisations (such as developers, city managements, governments). I build on these experiences, aiming, as Massey proposes in her essay *Negotiating Disciplinary Boundaries*, to give attention to the 'relations between the disciplines' (1999:8). Employing methods and literature from different disciplines supports an investigation into how public spaces are planned, constructed and used together within a social framework of how public spaces are produced through daily interactions. A combination of observation, interviews,[2] document surveys and visual analysis has enabled me to access a diverse range of accounts of public spaces as they are planned, envisioned and used. The research approach also attempts, as Mitchell Duneier undertook in his ethnography, *Sidewalk*, to 'make links between the micro and the macro' relationships that inform these public spaces (1999:344). The analysis extends beyond the specific spaces of the market, canal-side and civic square to follow decision makers of central government, Greater London Authority (GLA), global private investors and international corporations that drive urban regeneration in London. I also reveal stories of people, organisations and material spaces that are bound up with this marginalised market at Elephant and Castle, tightly maintained waterside at Paddington and highly managed civic space of Trafalgar Square.

I ask three questions: first, I am concerned with how masterplanning mediates between scales of political and economic strategies, management practices and everyday use. I highlight connections between ways of making that produce specific architectural and social forms of public space. Across the three cases investigated, I heard accounts of how policies are set out by politicians to facilitate neighbourhood-scale masterplans; how, as buildings are demolished, daily market operations are threatened; and how, as regulations are rewritten during redevelopments, new programmes of events are curated. I examine relations between the projects designed and the spaces enacted, how they came together and how they were contested. I compose socio-spatial narratives of each site that highlight relationships between organisations with power to impact large geographies and populations contrasting with smaller-scale groups and individuals seeking opportunities to use and influence public spaces.

Second, I ask how planning and masterplanning processes, with public spaces at their core, produce differential benefits and represent interests in uneven ways. I examine disparities in public space relationships, such as financial means, political power and rights of ownership over land that are leveraged when public spaces are reconfigured through planning and reinforced through their continued management. In each case, the building out of the masterplans undermines the continuation of certain activities and compromises future opportunities in the new open spaces. New management structures are also established through the redevelopment process in the form of BIDs, estate managers and private security teams enforcing codes of use and access. I recognise that public spaces are both the process and result of contested and unequal social relations, conflicts that are played out in open spaces, in the offices of planners and developers, in online forums and occasionally in legal challenges.

Third, I question how specific commonalities and differences across the sites reveal general insights into large-scale developments. I identify specific conditions and circumstances that have informed the public space narratives. In particular, I question how differences across the cases of masterplanning frame nuanced accounts of redevelopment: how the insecure future for the businesses and retailers at Elephant and Castle was caused by the opaque negotiations and the uncertain progress of the redevelopment; how decisions to exclude homeless people from Paddington Basin, both evicted from the development area and excluded from the neighbourhood as nearby homeless hostels were renovated into hotels, was informed by the priorities of the developers and BID managers; and how laws such as the 1999 Greater London Authority Act informed the intense schedule of events in Trafalgar Square that undermined smaller and more spontaneous public actions.

By exploring these questions, I identify competing ambitions for producing public spaces: made and managed in architectural terms by teams of developers, architects, politicians and planners (*spatial forms*); curated and shared through artistic renderings and a proliferation of media, such as newspapers, photographs, films and online accounts (*visual images*); and occupied, appropriated, used and reconfigured through diverse activities and events (*social interactions*).

Outline of chapters

At the core of the book is an investigation into intersecting accounts of public spaces. In Chapter 1, *Social and spatial relations* I examine concepts and histories of how urban spaces, and in particular public spaces in London, have been made and remade. I bring together discourses from across architectural and sociological disciplines that offer contexts for, and clues to, the issues of how public spaces in the three London sites are constantly in the process of being made through material, visual and social interactions. The chapter is structured with the aim of understanding the relations between scales of global competition that inform the development of the cases, as well as reviewing research focused on architecturally designed spaces and close-up scales of everyday actions.

The investigation of Elephant and Castle Market is discussed in Chapter 2, *Making and taking*. I examine how, despite a masterplan that envelops large swathes of Elephant and Castle, planning processes have progressed in a piecemeal and incremental manner. I examine scales of production, from the daily making and dismantling of market stalls to the comprehensive planning of the area spanning the last hundred years. I also question whether masterplanning strategies are reliant on the need to take space, where the problems caused by physical appropriation are compounded by drawn out periods of talking-down the everyday places of this marginalised South London neighbourhood. Chapter 3, *Place as property*, examines the sites of the *Paddington Waterside* masterplan—a process where a developer consortium and a BID has created a relentless commodification of space and claimed a certain making of places and public spaces. I argue that the business-oriented development results in a seamless, polished and over-written public realm that privileges people moving through the area rather than activities that may require a longer-term presence. I recognise the objectives and actions of state agencies handing over public assets to be managed by private concerns and an aggressive expansion by private interests to selectively claim and package spaces, images and narratives beyond their spatial or legal bounds. Chapter 4, *Ornaments and images*, sets out the remaking of Trafalgar Square by the GLA to establish a rhythm of globally visible events and practices in creating distinctive visual narratives of London. I question the visual and architectural emphasis, the civic presence of the square, its role as a public space and an intensity of maintenance. I highlight the ambitions for public spaces as scenes with which individuals and organisations wish to be associated. The chapter discusses schedules of international events that involve the fencing-off of the square and, through the occupation of the space by crowds of tourists and protestors, the identification of the square as a distinctly ornamental and spectacular setting. I then reveal how the conflicting daily routines and large-scale events and gatherings are managed and presented in and to the public.

Chapter 5, *Approaches to public spaces*, brings together the conceptual and analytical issues highlighted in the three preceding chapters. I discuss the overlapping, common narratives, revealing public spaces constituted as *spatial forms, visual images* and *social interactions*. I explain how researching these three public spaces has highlighted common concerns and conflicting imperatives that contribute to particular understandings of public spaces across London. I discuss unequal relationships between contested ways of realising public spaces and I critically question public spaces produced in order to realise economic ambitions. The final chapter presents conclusions from the research and focuses on the imposition of masterplans, aspirations for visual images and spatial forms and the realisation of large-scale strategies. I demonstrate how uneven distributions of political power, economic means, cultural reputations and physical presence influences the ability of people to claim, access, assemble in and even withdraw from the formation of these masterplanned public spaces. I go on to discuss three considerations for how public spaces could be reconceived: as socio-spatial relations of public spaces; as continual processes of making; and, in terms of understandings of 'publicness' focused on the inclusiveness of making rather than access to public spaces as products. In the epilogue, I set out three propositions for public spaces to advocate for an approach to achieving greater transparency through more justly and inclusively negotiating the future production of public spaces in London.

Notes

1. The three sites of focus in this study are the public spaces of Elephant and Castle Market, Paddington Basin and Trafalgar Square. They are situated within three cases of masterplanned redevelopment, namely *Elephant and Castle Regeneration*, *Paddington Waterside* and *World Squares for All*. Descriptions of 'Elephant and Castle', refer to the area or neighbourhood of Elephant and Castle. The term 'Paddington Basin' refers to the waterfront spaces from Little Venice to Paddington Basin, including the open spaces contiguous with it which form the *Paddington Waterside* masterplan (Paddington Basin is described by the Canal and River Trust as the canal basin approximately 500 metres south of Little Venice and it is also the name used by a previous development operation, now called *Merchant Square*). The name 'Trafalgar Square' is used to describe the entire space enclosed by the National Gallery and the buildings on the other three sides of the square.
2. All names of interviewees throughout the book have been anonymised, except for that of Deputy Prime Minister John Prescott who during the re-planning of Trafalgar Square was Secretary of State for the Environment, Transport and the Regions and responsible for planning. The high-profile role that Deputy Prime Minister John Prescott had during this period makes it impractical to anonymise his account.

1

SOCIAL AND SPATIAL RELATIONS

Over the last thirty years London has undergone an era of pronounced redevelopment where commercial speculation and public interests competed and where small-scale concerns have been overlooked. Approaches to transforming public spaces that were tested in North American and European cities in the 1980s have been adopted in varying ways in London—masterplanning processes that brought together government agendas with international investment. In this chapter I explore such reconfigurations of urban space within national and global contexts, processes that inform both practices of public space and notions of the public sphere. I highlight the intersection of spatial and visual approaches with social conceptions of how London's public spaces are made and remade. I discuss research and practice that inform the architectural configuration of public spaces (from urban policy to architectural theory) with discourses around their everyday remaking (from social sciences). I afford particular focus to conflicting ways that public spaces are made, highlighting unequal relations between global agendas and smaller-scale routines (Gospodini 2002; Low and Smith 2006; Mitchell 2003; Shane 1995, 2005; Sorkin 1992; Zukin 1995). By examining architecture, landscape architecture and urban design research alongside texts by sociologists, geographers and anthropologists, I explore relations between government policies, developer ambitions, spatial forms and daily activities.

In this chapter, I discuss definitions of public space found in the public spaces of Elephant and Castle Market, Paddington Basin and Trafalgar Square, extending from spatial forms (see Carr et al. 1992) to more social relations (see Massey 2005). I focus on the timeframes of development of the three cases of masterplanning (between 1998–2018); this is a period in London when renewed policy directions and highly charged economic agendas facilitated large-scale redevelopments. Exploring broader geographical contexts, I examine models of private

DOI: 10.4324/9781003248187-2

development and management that have transformed North American public spaces and that reflect both practices of regeneration and economic conditions for development found in London (Sorkin 1992; Mitchell 2003; Low and Smith 2006). I also explore research that describes state-led approaches to designing public spaces across Europe, strategic interventions that enhance the ability of European cities to compete for large-scale private investment (Degan 2008; Fainstein 2010; Gospodini 2002).

The chapter is structured in four sections and reflects, as Neil Smith and Setha Low describe in *The Politics of Public Space* (2006), '[how] the scale of public space and the public sphere is socially produced' (2006:7). I describe scales of producing public spaces that are identified in the three cases, highlighting relations of power from national policy and international corporations to community initiatives and individual actions. The first section intersects global agendas, national policies and metropolitan strategies for public space. In the second section, I analyse processes of masterplanned redevelopment, demonstrating how spatial forms are defined by urban designers for developers and local authorities and how their neighbourhood scales correspond to public realm management mechanisms. In the third section I focus on the architectural dimension of individual public space projects that are designed as scenic settings, constructed as spatial products and used as social spaces. In the fourth section I discuss scales of face-to-face interactions where people gather in, repurpose and regulate public spaces and how these actions can be understood in terms of both spatialised public spheres and constantly reproduced public spaces.

Policies and strategies for public space

The beginning of the masterplanning processes of *Elephant and Castle Regeneration* (1999 and 2004), *Paddington Waterside* (1998) and the *World Squares for All* (1996) are marked by significant changes in planning and development policy in the UK. At the end of the twentieth century, the recently elected New Labour government (1997) published *Towards an Urban Renaissance* (1999), a document authored by a newly formed Urban Task Force that set out conditions for the regeneration of British cities. *Towards an Urban Renaissance* was commissioned by Deputy Prime Minister John Prescott and was led by the architect Richard Rogers. The report emphasises the need for design-led regeneration of cities across the UK. It features a foreword by Pasqual Maragall, the former Mayor of Barcelona who had presided over his city's winning and hosting of the Olympic Games in 1992. The authors present international case studies of what they consider successful public spaces. They emphasise the importance of achieving a 'high-quality urban product by creating compact urban developments' (1999:11). The report proposes that developments include networks of public spaces composed of streets, squares and parks, forming a public realm managed by the public sector.

Towards an Urban Renaissance reflects many policy and design issues that Rogers had identified, with the politician Mark Fisher, in *A New London* (1992). Rogers and Fisher anticipate the establishment of the Greater London Authority (GLA) by calling for a 'strategic planning body' that could 'experiment with land taxes, road pricing, tourism taxes and business rates' (1992:xxxii). Following the publication of *Towards an Urban Renaissance*, the Urban Task Force and the government's Commission for Architecture and the Built Environment (CABE) authored further reports encouraging the regeneration of UK cities. Publications such as *The Manifesto for Better Public Spaces* (CABE 2004) and *Towards a Strong Urban Renaissance* (2005) greatly influenced a climate of design-led masterplanning and government-facilitated urban redevelopment for over a decade. The foundation of these organisations and their publications, Ben Campkin explains in *Remaking London*, 'marked a feverish new appetite for the regeneration of cities, and one continued by successive governments and local authorities ever since' (2013:2).

This focus on urban redevelopment emerged from national contexts of previous Thatcher-led Conservative government policies and the influence of international strategies for the development of new public spaces. In the 1980s Thatcher's government abolished the Greater London Council (GLC), London's metropolitan government led by Ken Livingstone, and began to facilitate specific commercial developments, such as London's Docklands. Rob Imrie and Mike Raco explain in *Urban Renaissance?* (2003:3):

> Regeneration, Thatcher-style, was characterised by the use of public subsidies, tax breaks, and the reduction in planning and other regulatory controls as a mechanism to create a context to encourage corporate capital to invest in cities.

Imrie and Raco highlight a government policy that claimed that a 'trickle-down' of wealth from capital investments would reach local communities (2003:11). They explain that significant criticism of the government's approach pointed to widening inequalities and increased poverty in cities. The resultant fragmentation of districts created enclaves of intense development juxtaposed with areas of neglect (see Shane 1995). This reliance on free-market approaches to development in the 1980s and early 1990s also led to new forms of tightly managed, commercially focused public spaces. Developments such as Canary Wharf and Broadgate established a highly securitised public realm around their estates of commercial buildings. In response to such developments, Rogers and Fisher claim that the public realm of London was being privatised. Emphasising the need for local authority control of public spaces, and criticising the increased presence of private interests, they state: 'there is no revival of public spaces until these issues are tackled' (1992:111).

During the 1980s two contrasting public space strategies were employed in several cities in North America and across Europe. On the one hand, the involvement of private interests and commercially led development of public spaces was embraced in cities such as New York. In contrast, some European cities, most notably Barcelona, embarked on significant government-led investments into waterfronts, plazas and neighbourhood parks. These two approaches to planning were later combined in varying ways in London's Urban Renaissance under the New Labour government. In North America, as metropolitan governments witnessed periods of deindustrialisation after the 1950s, decline in urban populations and limited financial means, they attempted to identify alternative development and management mechanisms for their public and private realms. From the 1960s, and intensifying in the 1980s, varying forms of involvement by private interests in the design, construction and management of urban areas were accepted by local governments, including: Business Improvement Districts (BIDs), park conservancies, privately owned public spaces, bonus plazas, suburban shopping malls, festival districts and gated campuses. Architect and writer Michael Sorkin brought together critics of these private strategies (including Crawford, Smith, Davis and Boyer) in *Variations on a Theme Park: The New American City and The End of Public Space* (1992). Further critiques of how cities were being redeveloped ranged from the deindustrialisation and the re-colonisation of abandoned cities (Zukin 1991) to their gentrification (Smith 1992, 1996), 'Disneyfication' (Zukin 1991; Sorkin 1992) and militarisation (Davis 1990; Mitchell 2003).

In contrast to strategies that had been conceived, tested and employed widely in North American cities, state-led investment into the public realm was prioritised in Barcelona in the 1980s. Emerging from an era of isolation and underinvestment during the Franco era, and witnessing decline in its traditional industries, the metropolitan government of Barcelona commissioned new public spaces and infrastructure as catalysts for economic development and as a focus for attracting inward investment (see Poynter 2009; Degan 2008). In *Sensing Cities: Regenerating Public Life in Barcelona and Manchester* (2008), Monica Degan writes:

> Creating more public spaces for collective use in areas such as El Raval [an historic neighbourhood in Barcelona] has to be understood as coming from the strong civic ideals that informed the first years of democratic planning.
> *(2008:96)*

This approach to regeneration, which prioritised state-initiated design and realisation of new public spaces, was closely observed and followed by other countries and cities who sent delegations to Barcelona to understand their success. Rogers writes in the introduction to *Towards an Urban Renaissance*:

> What we learnt from these visits is that regeneration has to be design-led. But to be sustainable, regeneration also has to be placed within its economic and social context.
> *(1999:7)*

However, without its own metropolitan government from 1986 to 2000, London was unable to coordinate redevelopment as New York or Barcelona had achieved. Although London underwent its own deindustrialisation and the abandonment of its waterfronts due to transformations in global competition, it was less well equipped to respond with coordinated urban strategies for its boroughs, neighbourhoods and public spaces.

But, by the end of the twentieth century London had engineered its own renaissance through bringing together Barcelona's model of state-initiated public space-led regeneration with commercially focused funding mechanisms that had been pioneered in North America. In *The Just City*, Susan Fainstein states: 'Within the history of redevelopment policy London represents an intermediate case between the New York model and that of the continental European cities' (2010:113). She explains that while London has had an 'activist and redistributional public sector' it also uses 'privatization, public subsidies, and deregulation to promote property speculation and entrepreneurship' (2010:113). The establishment of a new metropolitan government (GLA) and an elected Mayor for London facilitated high-profile initiatives, such as the *100 Public Spaces Programme*, the *World Squares for All* masterplan, of which Trafalgar Square was a part, and support for London's bid to host the 2012 Olympic Games. Although the GLA 'drew on lessons learnt from the free-market disregard for social policy and social issues in the regeneration of the London Dockland's' during the 1980s as well as the 'weak trickle-down effects of the Docklands' regeneration to the adjacent parts of the East End' (Imrie et al. 2009:14), the GLA still needed to engage with private interests to fund many of their initiatives.

The New Labour government did not reject the previous market-led approaches to development—instead they encouraged a closer association between private capital, commercial interests and the public sector. The Urban Task Force set out that the 'public sector must act as the custodian of the public realm' while simultaneously advocating the adoption of new financial instruments and incentives to encourage private investment in public space (1999:56). CABE's *Manifesto for Better Public Spaces* reflects the Urban Task Force by emphasising the need for 'coordinated funding' for public spaces 'from both the public and private sectors' (2004:9). However, as there were limited funds to provide state-led public spaces as achieved at Trafalgar Square, and to fulfil the ambitions of Mayor Livingstone's *100 Public Spaces Programme*, local authorities relied on developer contributions to provide design, construction and ongoing maintenance for new public spaces (see *Elephant and Castle Regeneration* and *Paddington Waterside*). The situation in London reflected that of other cities that over the previous decade had become dependent on private investments to facilitate regeneration. Writing in *City Builders* (1994) before the election of New Labour and the subsequent implementation of new development policies, Fainstein describes:

> The quandary for local political officials is that they must depend on the private sector to finance most economic expansion, and they have only very limited tools for attracting expansion to their jurisdictions.
>
> *(1994:2)*

With the election of the Conservative party and David Cameron as Prime Minister (2010–2016), funding for organisations like CABE and the Design Council (2011) were severely cut and agencies such as British Waterways (that later became the Canal and River Trust) were privatised to become independent trusts (2012). Similarly, public space investment was curtailed as Boris Johnson became Mayor (2008–2016), resulting in the later phases of the *World Squares for All* masterplan remaining incomplete. In contrast, masterplans that involved commercial developers continued to be facilitated, such as *Elephant and Castle Regeneration* and *Paddington Waterside*, which were prioritised as Opportunity Areas (www.london.gov.uk 2016). Despite contrasting policies under successive governments, these approaches to development are all 'part of a broader socio-political process in London which places urban regeneration at the fulcrum of the capital's economic competitiveness' (Imrie et al. 2009:5). Imrie, Lees and Raco argue that:

> regeneration is being 'put to work' by politicians as part of a strategy to remove obstacles to economic growth and to create the social and physical infrastructure required to compete for inward investment.
>
> *(2009:5)*

These policies 'conceive of regeneration as closely entwined with globalisation' (2009:6) with new and refashioned public spaces being key to inserting and maintaining London in these global city relations. In *European Cities in Competition and the New 'Uses' of Urban Design*, Aspa Gospodini describes the way that the global system is 'increasing competition among cities to upgrade their status' (2002:60). She writes that the 'development prospects' for European cities are considered partly due to the 'high quality of urban environment' (2002:60). Political ambitions were heightened in 1990s London as architects and politicians witnessed Barcelona prioritising the creation of hundreds of new public spaces, hosting the 1992 Olympics and raising its international status (Urban Task Force 1999). From the late 1990s these ambitions began to be expressed in the remaking of public spaces in London.

The refashioning of urban space in London through the lens of global competition has been employed to spur economic regeneration: the redesign of Trafalgar Square can be seen to empower London's global standing through its imagery and events, while the regeneration of Elephant and Castle and Paddington Basin drive inward investment through residential sales and commercial lettings. However, these selectively redeveloped areas are claimed to also increase fragmentation across London. In *Urban Design Since 1945*, David Grahame Shane contends that 'London emerged as the exemplary fragmented metropolis in the 1980s' (2011:27). Building on earlier writing describing London's 'enclaves of hyper-development' and contrasting 'enclaves of disinvestment' (1995:65) Shane suggests that increasingly market-led approaches, as pioneered in London in the Docklands, led to 'local areas of architectural and urban design control in urban villages' (2011:25). Selectively chosen Opportunity Areas have further contributed to this fragmentation, facilitating the economic development of strategic areas that contrast distinctly from locations left lacking investment. Despite the Urban Task Force recognising problems of fragmentation in UK cities (1999:50) the GLA has continued to 'provide[s] encouragement, support and leadership' for Opportunity Areas to address concerns that 'London has limited opportunities for accommodating large scale development' (www.london.gov.uk 2016).

Masterplanning districts with public space

Masterplans are the mechanisms through which government policies and many commercial objectives for development are focused and the way that *Elephant and Castle Regeneration*, *Paddington Waterside* and *World Squares for All* are architecturally directed. These top-down approaches to planning, commissioned by commercial developers or state agencies, and undertaken by urban designers, give direction and guidance for long-term, large-scale development. The main goals of these masterplans have been to organise and, if necessary, gain planning approval for spatial development. Broader urban design ambitions are to make visible urban change to compete with other developments, districts and cities to attract investment. Gospodini describes how this approach contrasts with historic urban development:

> While for centuries the quality of the urban environment has been an outcome of economic growth of cities, nowadays the quality of urban space has become a prerequisite for the economic development of cities; and urban design has undertaken an enhanced new role as a means of economic development.
>
> *(2002:60)*

Imrie et al. (2009), Shane (2011) and Gospodini (2002) claim that this form of competitive masterplanning in London began with the Docklands. While subsequent London developments have departed from creating similar highly commercial, tightly controlled and privatised estates, such economic imperatives remain in the foreground of government and developer ambitions. Beyond the nuanced conditions addressed by *Elephant and Castle Regeneration*, *Paddington Waterside* and the *World Squares for All* exist the priorities of organisations commissioning masterplans to architecturally transform sites of sufficient scale to attract investment and returns.

The spatial scale of the masterplans encompasses large areas of London that include networks of spaces and buildings. Their scale incorporates neighbourhood-sized areas of the city, representing what Kevin Lynch terms a 'district' (1960:66). Lynch considers districts as sections of cities that 'the observer mentally enters "inside of" with common interior characteristics and exterior references' (1960:47). As London has never been structured around a single and dominant city plan, unlike Barcelona whose extension was meticulously planned by Cerda (1859) or New York City with its grids of streets and avenues (1811), London's aggregation of districts is how many people perceive and experience the city. *Legible London: Yellow Book*, published by Transport for London, explains that 'London has no structured delineation' but is instead composed of 'a rich collection of neighbourhoods and boroughs' (2007:13). Under the oversight of the GLA, local authorities administer planning processes across their boroughs. This is highlighted by Fainstein who claims that 'despite the creation of the Greater London Authority and the Assembly the metropolis remains decentralized to the thirty-three local authorities' which 'continue to be the main decision makers regarding services and the specific forms that development takes' (2010:136).

Masterplanning across London's boroughs focuses on defining districts, neighbourhoods and urban villages. As Shane demonstrates in his field analysis of central London, featured in Colin Rowe and Fred Koetter's *Collage City* (1978:114), historic masterplanned estates include eighteenth and nineteenth century districts such as Covent Garden, Belgravia, Mayfair and Chelsea. Shane explains that London has had a 'long fractal tradition of developing large, single-landowner enclaves' (2011:27), including the arrangement of the historic docks and the contemporary Docklands, which mirror the power of property ownership found in the 'Great Estates'. These sites of urban development, which evidence contrasting architectural typologies of public space, also include post-war government-built housing estates, such as the Heygate Estate (Elephant and Castle), the Golden Lane Estate (City of London) and Robin Hood Gardens (Poplar). Since the 1980s, this district scale of masterplanned development has been undertaken in free market conditions, realising new estates through masterplans such as Paddington Waterside, King's Cross Central, Canary Wharf and Broadgate.

Masterplans are an urban design tool to spatially and economically restructure neighbourhoods as well as defining new districts. The first 'objective' of urban design, as defined by the former UK government's guidance in *By Design: Urban design in the planning system: towards better practice*, is to 'promote character' (DETR and CABE 2000:15). This objective resonates with Lynch's notion of 'districts' which he describes as being identifiable through 'some common character' (1960:66). Masterplanning creates districts with strong identities, often contrasting with less well-defined adjacent urban areas. Lynch argues how districts can be understood through their 'thematic continuities' which include their 'texture, space, form, detail, symbol, building type, use, activity, inhabitants, degree of maintenance, topography' (1960:67). Similarly, the DETR and CABE list six other objectives, including: continuity and enclosure, quality of the public realm, ease of movement, legibility, adaptability and diversity (2000:15). These terms are promoted as the means through which masterplans can be realised and the ways that, once built, public spaces can become read as architectural forms. Through reinforcing 'thematic continuities' urban designers can realise specific ambitions for their clients, such as creating accessible public spaces, symbolic landscapes and exclusive districts. The image of large parts of cities therefore becomes transformed through masterplanning—as Lynch reminds us, 'most people structure their city [image]' through the identity of the district (1960:47).

A range of strategic instruments of development can be employed when designing urban areas. Beyond masterplan drawings, which represent the spatial qualities of the proposed district, masterplans frequently include design guidelines, design codes, management plans and public space regulations. However, these devices are insufficient on their own to control the developments, so further legislation and planning mechanisms are pursued by developers and local authorities. In addition to the tools available through masterplanning, which can redefine entire neighbourhoods, additional development approaches, such as BIDs and CPOs (Compulsory Purchase Orders), are employed. Low describes 'physical tactics' of enclosure, eviction, securitisation and acquisition that are enhanced by 'legal and economic strategies' (2006:83). The Urban Task Force writes that both 'compulsory purchase and the threat of compulsory purchase are powerful tools for securing urban regeneration' which are part of a 'package of tools and measures' (1999:228). CPOs are mechanisms available to developers and local authorities who need to obtain the ownership of land for developments, which they can argue will promote or improve economic, social or environmental well-being (www.legislation.gov.uk 2016). We will see in Chapter 2, *Making and taking: Elephant and Castle*, how CPO powers were proposed to be used by local councillors to encourage the former shopping centre owner to demolish rather than renovate their property. Once land has been obtained for these developments, districts are more easily remade architecturally. In the three cases we witness these strategies differentially: through the threat of CPOs at Elephant and Castle (Planning and Compulsory Purchase Act 2004); in the establishment of the BID (The Business Improvement District (England) Regulations 2004) and the outsourcing of the public realm management to Broadgate Estates at Paddington Basin; and in the new regulations put in place at Trafalgar Square under the Greater London Authority Act 1999.

Where land cannot be purchased, or it is not in the interests of those with power to acquire it, organisations such as BIDs can be formed to exert greater control over urban districts. A proliferation of BIDs, commercially led organisations that oversee the management of areas of London's public spaces, has occurred in London since the passing of the Business Improvement Districts (England) Regulations in 2004. New districts have been established, defined through branded signage, cleaning, security and design of the public realm, reflecting the interests of businesses. We can recognise such operations in PaddingtonNow (the BID at Paddington formerly known as PaddingtonOn) that organises its visual and spatial structure, as the *Legible London Yellow Book* sets out, 'to suit their own purposes' (2007:14). In his essay *The Ends of Urban Design* (2011), Sorkin claims that the importance of BIDs 'has only grown as government has become increasingly enthralled by the model of the "public-private" partnership' (2011: 292). This can be problematic as the 'benefits of urban design (and maintenance) are directed to commercially driven players' (2011:292) facilitated through favourable Section 106 agreements, CPOs and BID mechanisms.

Masterplans are criticised as problematic devices in urban design. The large-scale, far-reaching processes and tools, which incorporate what Rowe and Koetter describe as 'total planning', 'total architecture' and 'total design' are closely associated with modernist city-plans that demonstrate insensitivity to their contexts (1978:86). Rather than single-use, mono-functional urban plans, Rowe and Koetter's critique of utopian and modernist approaches to the city instead advocate a collaging of elements, including public spaces of 'memorable streets' and 'splendid public terraces' (1978:152–164). Sorkin similarly recognises the problem with what he explains as 'the inherent dangers of giant, single-sourced plans' (2011:292) but he also questions alternative approaches premised on 'a suspicion of big plans [that] refuses, however provisionally, to sum up its parts' (2011:292). We see in many London masterplans that the need to consider the city spatially places architectural designers in a prominent role during planning discourses. This can lead to an over-emphasis on architectural frames of urban design with deliberations over form and style eclipsing other considerations. While Rowe and Koetter's critique was primarily aimed at large-scale modernist masterplans, the style-based forms of post-modern masterplanning also fail to address more socially oriented concerns of large-scale developments. For Sorkin, postmodern masterplans 'evoking styles of manufactured difference' fail to address previous modernist approaches that also lacked diversity. He states: 'Today's urban nightmare is the city in which the differences are simply architectural' (2011:373).

Fixing public space architecturally

Architectural public spaces are core elements of masterplans, providing shared spaces for people to gather in and to pass through. Long-established architectural typologies of public space, such as streets, squares, plazas, parks, amphitheatres, terraces and footpaths, dominate masterplans. In their multi-disciplinary research, *Public Space*, Stephen Carr et al. describe such public spaces (1992:3):

> These dynamic spaces are an essential counterpart to the more settled places of work and home life, providing the channels for movement, the nodes of communication, and the common grounds for play and relaxation.

Their investigation, which includes how public spaces are designed, built, used and managed, focuses on spatial frames of public space. Their description of architectural public spaces as the common ground on which people 'carry out' activities (1992:xi) closely reflects Michel de Certeau's description of tactics which are limited to 'play on and with a terrain imposed on it' by those with power (1984:37). These public spaces represent consistent spatial components that fit together with buildings, structures and urban networks, establishing the material qualities of masterplans. The Urban Task Force posits that the benefit of a 'spatial masterplan' is that it 'controls the relationship between buildings and public space' and that it 'shows how streets, squares and open spaces of a neighbourhood are connected' (1999:73). Masterplans are typically led by a single consultant team, such as Make Architects (who designed the area of the former Heygate Estate within *Elephant and Castle Regeneration*), Rogers, Stirk, Harbour and Partners (who designed the Merchant Square area of *Paddington Waterside*) and Foster and Partners (who led the *World Squares for All* masterplan), on behalf of those who own or control the land. The relationships between lead consultants and clients establish the appearance of confident and comprehensive proposals that can then be communicated to investors and local authorities.

Behind the perception of single-authored plans, approaches to realising public spaces as architectural forms vary from case to case. Carr et al. explain that 'public spaces are created and maintained by a lengthy and often complex process of interaction among sponsors, builders, managers, and users' (1992:292). They are critical of what they term a 'standard process' that maintains a narrow group of decision makers 'in the interests of central control, and presumed efficiency and economy' (ibid.). Clients engage closely with designers and other consultants to identify the issues in an urban site, redesign the space and contract builders for the construction. In *Design of Urban Space: An Inquiry into Socio-Spatial Process* (1996), Ali Madanipour explores broader relations between urban spaces and the practices through which they are produced. He uses the actions of urban design as a lens through which to examine contrasting scales of development, including 'macro-urban design and [a] micro-urban design' (1996:96). He writes that 'urban design deals with all scales of urban space' from the macro-scale design of masterplans to the micro-scale design of public spaces, and works across a gradient of scales of time. Through introducing the notion of time, Madanipour questions whether urban design should focus on the process or product. While he recognises that many architects and organisations are largely interested in the product of their work, he insists that urban design should be concerned with both 'the process of this shaping and the spaces it helps to create' (1996:105).

To design and create public spaces as architectural products, understanding relations of property is necessary. For Madanipour, 'land and property development is the vehicle through which the built environment is produced' and in order to 'understand the urban design process . . . it is essential to gain an understanding of the property development process' (1996:121). This assertion resonates with the developments at Elephant and Castle and Paddington Basin where transfers of control of property were fundamental to the realisation of projects. More specific to architectural forms of public space, Mitchell and Staeheli examine, in *The People's Property: Power, Politics and the Public* (2008), public space through an exploration of property. They find that 'property is a crucial part of the equation that creates public space and forms the public in its many configurations' (2008:128). They outline a 'practice of property' that includes the 'relations, regimes and struggles over what property is and how it is deployed' (ibid.). These relations are expressed through exchanges and appropriations of land during masterplanning, while new forms of property can be identified once physical public spaces are built and the rules of access are contested. Lynn Staeheli and Don

Mitchell describe how 'property ownership is a powerful tool in the regulation of space and, thereby, the public' (2008:xxiv). In London, this leads to competition between how public spaces are regulated, how they are intended to be used and rights over public and privately owned spaces.

Land has been exchanged to facilitate the spatial development of Elephant and Castle Market, Paddington Basin and Trafalgar Square. As Staeheli and Mitchell show in their research: 'property is a crucial set of relationships that structure the roles, function, and nature of public space as space' (2008:xx). However, changes to property ownerships are not the only mechanism available to organisations redeveloping and managing public spaces. Staeheli and Mitchell claim BIDs provide other means to regulate public space and transform the relationships of property (2008:151). In addition to the establishment of the BID at Paddington Basin, the transfer of control from the Crown to the GLA at Trafalgar Square provided new regulations without necessitating changes to the ownership of the space. In contrast to the privatisation of some public infrastructures, the regulatory transformations of public spaces have resulted in a range of privatisations that create fragmented and layered responsibilities of ownerships, leases and management.

The implication of defining public spaces as property is that it limits terms of public spaces to how they are managed and used. Control of public spaces allow owners, managers and other authorities to exclude people and activities. When these controls are expanded across masterplanned areas, large swathes of London become conditioned by small groups of individuals and organisations with power. Recognising that public space in London continues to fragment in terms of ownership and management (see Carmona et al. 2008), the GLA commissioned a report into *Public Life in Private Hands: Managing London's Public Space* (2011). As many public spaces in London are privately owned, including Elephant and Castle Market which was built prior to highly documented privatisations since the 1970s (Minton 2006, 2012), the relationship between the management and use of public spaces becomes more pertinent. The use of public spaces can be associated with what Kevin Lynch and Lloyd Rodwin describe, in *A Theory of Urban Form*, as 'traditional "land-use" categories of the planning field' (1958:204). The use of buildings, from factories to residential homes, is restricted by planning regulations that aim to provide a certain urban order and arrangement. Referring to cities, Lynch and Rodwin recognise that 'the pattern of activities and the physical pattern are often surprisingly independent of each other' (1958:204). But I would suggest strong relations can be identified between the spatial conditions of public spaces and the actions that occur. Across the three sites, the public spaces are appropriated for contrasting purposes; the physical geographies and conditions of spaces, materials and infrastructure provide for or deter certain activities. However, as is also revealed, spatial design is insufficient on its own to control or determine all uses. In such situations, as Lynch and Rodwin identify in practices of urban design, planners attempt to 'change the activity through the physical change' and by employing 'negative prohibitions' of regulations and laws (1958:204).

As public spaces are designed as spatial products, emphasis on the views of and from the architectural projects predominate. This issue is explored by Mitchell (1997) who questions relations between public space and visual aesthetics. He poses the question: 'landscape or public space?' (1997:322), claiming that the historic Anglo-Saxon approach to landscape, extending from picturesque scenes painted to urban public spaces designed, is a 'way of seeing the world, [one] in which order and control over surroundings takes precedence over the messy realities of everyday life' (Mitchell 1997:323). We can recognise Mitchell's concerns for an aestheticisation of public space when reflecting on political agendas, such as Mayor Johnson's *Manifesto for Public Space* that advocates: 'A beautiful city where the spaces between the buildings can inspire, excite and delight visitors and Londoners alike' (GLA 2009:2). Also emphasising aesthetic qualities of space, *Towards an Urban Renaissance* describes the spatial masterplan as a 'sophisticated visual model' (Urban Task Force 1999:73). Mitchell incorporates a critique of landscape set out by Henri Lefebvre, in *The Production of Space*, where Lefebvre is concerned with 'the power of landscape' that offers 'an already clarified picture' (1991:189). Lefebvre suggests that landscape, when embraced in aesthetic terms, gives the illusion of control: in cases of architectural design, developers, landowners and local authorities are led to believe that visually dominated urban design can provide control over spaces and their activities. Transforming public spaces through these terms denies many less attractive, everyday interactions in public spaces or the possibility to reconfigure them in less photogenic terms. Visual images do not merely represent, but are also appropriated to reinforce spatial priorities. In *The Cultures of Cities*, Sharon Zukin states that 'the power to impose a coherent vision of a space enables a group to claim that space' (1995:279), leading to other people and uses being excluded:

> The look and feel of cities reflect decisions about what—and who—should be visible and what should not, on concepts of order and disorder, and on uses of aesthetic power.
>
> *(Zukin 1995:7)*

In the three masterplanning processes, claims over public spaces are expressed strongly by policy makers, designers, developers and managers as they transform how spaces look, decide who is afforded access and define what they are permitted to do once inside. As Mitchell argues, through appropriating landscape techniques for making public space 'propertied classes express "possession" of the land, and their control over the social relations within it' (1997:323).

Strong visual imagery of public spaces can also obscure subtexts of control and ideology. James Corner, the landscape architect who designed the South Park Plaza of London's Queen Elizabeth Olympic Park as well as New York's High Line, writes in the essay 'Eidetic Operations and New Landscapes', that landscape provides 'a set of instruments to not only describe the world but to condition and control it' (1999:155). He also raises concerns for designing landscapes as spatial objects:

> A too narrow concern for landscape as object (whether as formal composition or as quantifiable resource) overlooks the ideological, estranging, and aestheticizing effects of detaching the subject from the complex realities of participating in the world.
>
> *(1999:156)*

As public spaces are defined as architectural products, their visual qualities become important. The framing of public spaces in visual images extends from architectural representations, developed by designers to communicate conditions of spaces and social lives contained within, to photographs, films and images taken of spaces once they are built. Design images are used to reinforce idealised scenes of proposed public spaces while images of realised public spaces document how these spaces and the activities in them are controlled. Zukin states: 'As in architecture, the visual image that designers create has been integrated into the landscape of power' (1991:48). In these instances, where it becomes necessary to limit certain uses and users, public spaces designed as landscapes are objectified as visually aesthetic forms.

We have seen that, despite traditions of public spaces framed as spatial forms, ambitions for public spaces vary. Carr et al. propose that 'the primary motives for making and remaking public spaces should be viewed against the changing panorama of public life' (1992:10). They set out 'public welfare, visual enhancement, environmental enhancement and economic development' as the goals of governments, corporate developers and managers for producing public spaces. In this research I inquire beyond the intentions stated in the development masterplans, to understand the stories of different people involved and their divergent goals for public spaces as spaces, images and interactions: the need for spatial infrastructures to enable events and connectivity is fulfilled through remaking physical public spaces; the desire for visual images of public spaces is met through hosting and producing scenic settings; while sites of interaction are realised as people meet and gather in Elephant and Castle, Paddington Basin and Trafalgar Square. Carr et al. also warn of 'not always stated' ambitions for making public spaces that may change in response to political and economic demands (1992:10). Gospodini (2002:61) elaborates on such economic priorities:

> In order to secure development and growth, 'localities' or individual cities now have to offer even more inducements to capital, whether a refashioning of the city's economic attractiveness (e.g. tax abatements, property and transport facilities) or alterations to the city's image through manipulation of its physical form and/or its soft infrastructure (e.g. cultural and leisure amenities).

Public spaces of interactions

How sites are occupied and appropriated, and how they facilitate interactions, allows individuals and groups to define public spaces in spatial and social ways. In Elephant and Castle Market, Paddington Basin and Trafalgar Square activities range from the routine to the spectacular, from conversations in street markets to erecting structures for events and from photographing friends to recording beautiful settings. Although not all interactions within public spaces constitute public actions, to be present in public or to simply use public spaces is an essential part of living in a city like London. In describing rights to public space Carr et al. see 'spatial rights' as extending beyond mere access (1992:137). They claim that the 'rights to use a public space and have a sense of control within it are basic and overarching requirements' (1992:137). However, these rights are not always explicitly stated in the design of public spaces or the regulations that provide for or restrict their use. At times, when regulations were enforced in Elephant and Castle Market, Paddington Basin and Trafalgar Square, rights to public spaces were undermined. In *The Right to the City: Social Justice and the Fight for Public Space* Mitchell states: 'what public space *is*—and *who* has the right to it—is rarely clear' (2003:5). Mitchell's research focuses on specific cases where he is concerned with 'the relationship between social activities and changes in public space law' and in particular, marginalised people, such as homeless people, who are excluded from certain urban spaces (2003:5). The following chapters discuss moments of contestation through accounts of homeless people, street vendors and teenagers transgressing site regulations, revealing that some groups' rights to use redeveloped public spaces are restricted.

The public nature of urban spaces is also frequently described in terms of accessibility. What we could term the *publicness* of space is often framed through 'rules of access' to a city's squares, streets, parks and plazas (Smith and Low 2006:3). In *Whose Public Space?* Madanipour argues: 'The key feature of public space . . . is its accessibility' (2010:8). Issues of access mainly settle on two interrelated questions of who is afforded access to these architectural spaces and what activities are permitted. In Elephant and Castle Market, Paddington Basin and Trafalgar Square there are regulations that correspond with the architectural transformations of the masterplans. Terms of access do not explicitly exclude individual people, but the regulations focus on descriptions of spatial forms and social actions that are prohibited. As revealed by Staeheli and Mitchell in public spaces in North America (2008), regulating against the use of sleeping equipment and unlicensed gatherings in London's public spaces is an effective way of denying the presence of homeless people and protestors. Reflecting Lynch's spatial right of 'presence' (1981), Carr et al. define three forms of 'access', including physical, visual and symbolic access (1992:150): physical access is the ability to enter a public space (1992:138); visual access is the ability to see into a public space from outside and is considered to be significant for public safety (1992:144); while symbolic access relates to the 'presence of cues' (1992:149) which can be read to indicate who is and who is not welcome. Cues include: signs within spaces that prohibit activities of skateboarding, cycling, unlicensed performances, vending, the use of sleeping equipment and feeding pigeons; the presence of physical obstructions, such as bollards and impediments to laying down; and security guards enforcing the rules. In Elephant and Castle, the stop-and-search techniques of the police made the plaza by the shopping centre a disputed territory for teenagers at night; in Paddington Basin individuals cycling along the edge of the canal are stopped by private security guards; while online and traditional media depict a rhythm of fenced-off events at Trafalgar Square, emphasising the GLA's priorities for one of its main public spaces.

Through the next three chapters, I discuss concerns for public spaces that are more or less accessible, who is allowed access and what people are permitted to do once inside. But I more explicitly examine the accessibility to ways of making public space, including questions of who is afforded what opportunities. Contesting historical descriptions of public space, Mitchell explains:

> public spaces were only public to the degree that they were *taken* and made public. Definitions of public space and "the public" are not universal and enduring; they are produced through constant struggle in the past and in the present.
> *(2003:142)*

Mitchell perceives public space similarly to Nancy Fraser's (1990) consideration of the public sphere—as a site of struggle. Contestations over space and the issues that they represent range from rights of presence to opportunities to be involved in a site's production. Staeheli and Mitchell write that 'being present in public space—making claims to and becoming visible in the streets, sidewalks, squares, and parks of the city—is a vital, necessary step in making claims *on* the public and *as part of* the public' (2008:xiv). Due to the demands on public spaces, from global economic forces, government agendas, masterplan controls and architectural styles, we can identify how certain behaviours are restricted. As landowners, developers, investors and governments increasingly have potential profits and reputations at stake in the production of public spaces, opportunities to claim these spaces by other individuals and organisations are increasingly challenged.

As Staeheli and Mitchell (2008) suggest, rather than mere access to physical spaces, the access or opportunity that individuals and groups have to remake public spaces can provide more nuanced clues to a site's publicness. In *For Space*, Doreen Massey suggests that, in addition to concerns associated with the privatisation of public space, 'we might address the question of the social relations which could construct any new, and better, notion of public space' (2005:153). Relations around making public spaces are of particular interest in this research and underscore the usefulness in bringing together architectural research focused on physical spaces and social science literatures that discuss publics and the public sphere. I argue that discourses around the public sphere and public space have progressed separately, a concern raised by Smith and Low (2006:5). They highlight what they perceive to be the weakness in existing public space literature, 'in the practical means of translation from the theories of political and cultural economy to the materiality of public space' (2006:6). Smith and Low question the 'separate domains' that the public sphere and public space occupy and the infrequency with which these discourses overlap, highlighting the potential of further 'investigating the spatiality of the public sphere' (2006:7).

Relations between public spaces and the public sphere are alluded to by Fraser in *Rethinking the Public Sphere: A Contribution to the Critique of Actually Existing Democracy* (1990). Critiquing Jürgen Habermas' conception of the public sphere (1962), Fraser states that the public sphere is 'a site of the production and circulation of discourses' (1990:57) where there is 'a plurality of competing publics' (1990:61). This perspective of the public sphere as 'a site' where publics are produced and fought over is reflected in Massey's three proposals for space. Massey states that first, 'space is a product of interrelations'; second, these relations are 'predicated upon the existence of plurality'; and third, that these spaces are 'always under construction' (2005:9). In the three sites, both Fraser's descriptions of publics and a public sphere, and Massey's proposals for space can be read. Public spaces are continually made architecturally and remade through social interactions in and across Elephant and Castle Market, Paddington Basin and Trafalgar Square. Accepting public space in these terms prioritises spatialising of emerging publics, conceives of public space beyond narrow architectural terms (of market plazas, canal towpaths and civic squares) and challenges restrictive regulations enforced by shopping centre managers, private security (and BIDs) and Heritage Wardens.

In addition to analysing the public nature of space, a stronger spatial understanding of the relations that form public space is suggested by Massey:

> For instituting democratic public spaces (and indeed the spaces of places more generally) necessitates operating with a concept of spatiality which keeps always under scrutiny the play of the social relations which construct them.
>
> *(2005:153)*

Through attending to the spatialising of relations that create public spaces, I question how people are differentially empowered or disenfranchised through redevelopment processes. Returning to issues of access to public space, who is included in making public spaces, how they can engage in masterplanned developments and the degree of change that they are able to inform are recurring questions. In an interview with *Shared Spaces* at the CCCB (Centre de Cultura Contemporània de Barcelona), Massey describes that 'public space is an arena, if you like, in which we have the opportunity of constructing a public' (www.publicspace.org 2013). Elaborating on this premise and accepting Massey's proposition that space is always being constructed (2005:9), could we also consider public spaces in terms of processes where people are ensured opportunities to reconfigure material public spaces? Studying relations between contrasting accounts of public space, from observation, interviews, documents and visual analysis, highlights differing opportunities for institutions, groups and individuals to participate, contribute and be a part of public spaces. I contend that what is often less considered when public spaces are focused on notions of policy, space, property, use and access, are opportunities to participate in the making and remaking of public spaces.

Conclusions

The structure of the chapter, from global and national relations to the scales of public space interactions, highlights the junctures of competing priorities and unequal opportunities in the making of public spaces. It demonstrates that 'urban space will inevitably reflect the values and aspirations of those who produced it' (Madanipour 1996:109). I have revealed contrasting perspectives, such as the fragmented conditions of London that are differentially addressed by critical theorists such as Shane (1995, 2011) as opposed to the government's Urban Task Force (1999) which addresses concerns for fragmentation through incorporating private interests. The chapter highlights the emphasis given by the New Labour government and the later Coalition and Conservative administrations on local authorities establishing relationships with the private sector in order to facilitate design-led redevelopment. Practices of masterplanning have achieved a new prominence despite continued criticism of such hierarchical 'top-down' spatial reorganisations of London. I have shown that despite these government priorities, an enthusiasm for urban regeneration in London is not new and that the development of masterplanned urban districts has been a key approach to achieving spatial transformations. The structure of the chapter also points to the significance of masterplans as mediatory mechanisms translating government policies and competition between cities into the physical and social fabric of London. We can conclude that while architectural forms of public space were afforded greater presence in policy and urban design during the period of masterplanning the three cases, the focus on economic agendas and private interests has left discourses around public spaces dominated by concerns for architectural form, ownership and management.

Rather than focusing on either architectural forms of public spaces or less tangible public spheres, I conclude that public spaces are gatherings of people (and associated institutions and organisations) around issues of concern within, and defining, physical, material geographies. This refers both to more political relations that contribute to the formation of publics as well as more architectural settings of the market, canal-side and civic square. As Smith and Low (2006) advocate, research that brings social relations and spatial conditions into proximity is particularly useful for understanding the contested dimensions of public spaces. Madanipour's definition (2010:1) of public spaces as 'accessible places developed through inclusive processes' opens up questions of how the design, management and governance of public spaces impacts on issues of access and inclusion. As I have demonstrated in this chapter, I give less emphasis to narrow frames of public space, whether from authors who critiqued the increased involvement of private interests in North American cities in the 1980s (see Sorkin 1992), research into subsequent privatisations in UK cities (see Minton 2012) or concerns regarding governance and management (Carmona et al. 2008). While the transfer of the management, and in some cases ownership, of public spaces into private hands is a concern, these changes to public spaces are only part of wider socio-spatial narratives.

I demonstrate in the following three chapters that public spaces are constantly remade by and inseparable from differing constellations of national policies, planning conversations, urban design projects, media descriptions, local histories, management decisions, scheduled events and daily actions. Building on Massey's propositions for space (2005:9), I find that the public spaces of Elephant and Castle Market, Paddington Basin and Trafalgar Square are constantly being made and remade through a range of intersecting actions, concerns and physical forms. I argue, that public space is simultaneously a *process* of making (that is focused in particular physical locations) and a constantly changing *product* of these interrelations (that derives from a wide range of geographies). I contend that to consider public space, the spatial forms cannot be uncoupled from the social relations from which they are formed. By examining the processes of making public spaces, unbalanced relations and contested public spheres—from which public spaces are constructed—can become more explicit. I build on this argument to claim that by investigating the relations between the processes (as policies, decisions and occupations) and spatial forms (as planned, designed, occupied and reconfigured) of public spaces it is possible to understand what is at stake as these sites are produced.

2

MAKING AND TAKING

'I know [Elephant and Castle] well. It always looks like something is going to happen and then nothing happens. . . . It always misses the cycle.'

—Sharon Peters, urban planner and resident of Paddington Basin (Interview 2012)

In South London's Elephant and Castle there has been a perpetual 'making and taking' of space. Plans for comprehensive urban regeneration have overshadowed the deteriorating architectural forms, such as the Elephant and Castle Market, and they have impacted the lives of the people who live, work and pass through the area. Strategic redevelopments have unfolded over the last century physically replanning the area and enabling ways of taking through acquiring, transferring and consolidating property. Historic attempts since the 1880s to replan Elephant and Castle provide a background for the latest appropriation of buildings and public spaces and the resultant displacement of people living and working in the area. Its proximity to central London, the considerable economic value of the 22-acre 'land bank' of the public housing Heygate Estate and the high-volume transport interchange have drawn public authority landowners together with private investors and commercial operators. These conditions create opportunities to realise large-scale strategic redevelopments provoking the proposition of new architectural visions while unsettling local people.

In this chapter I explore the Elephant and Castle Market in the context of the *Elephant and Castle Regeneration* masterplan. The redevelopment is the latest attempt at comprehensive change planned by the London Borough of Southwark and implemented in phases by private development partners, Lend Lease. By weaving between historic and contemporary periods of redevelopment I explore

DOI: 10.4324/9781003248187-3

FIGURE 2.1 The Elephant and Castle Market was squeezed into the public space of the moat that wrapped around the now demolished shopping centre.

(Photo by author)

the repeating narratives of developers, planners and architects reimagining the area. These accounts intersect with experiences of negotiating space between market traders, changing demographics of residents and businesses whose premises have been under threat. I examine how beneath the large-scale masterplanned regenerations, which involve demolishing and rebuilding large parts of the area from the 1950s to the 1970s and again since 2007, there are new private interests remaking the area's public spaces. I investigate histories of claiming, occupying, using and neglecting public spaces, all within the shadow of large-scale interests appropriating areas of Elephant and Castle during discontinuous periods of masterplanning, thus creating uncertainty and disadvantage for many local residents and businesses.

The investigation is focused on the public space of the low-cost market that, until it was demolished in 2021, was wedged between the shopping centre and the roadway in a narrow, sunken, publicly accessible open space called the 'moat'. The market operated between the annual extension of the market lease and daily rhythms of market stalls and between the incremental adaptation of individual buildings and the demolition of entire neighbourhoods. Each day, as the market opened, traders, shoppers and commuters took over the sunken space between the road and the shopping centre. Traders rented spaces in the low-cost market, erecting makeshift stalls, opened the repurposed shipping containers and set goods out across steel frame structures. As the market closed in the evening the plaza was cleared and cleaned. Large-scale transformations have been planned every few decades at Elephant and Castle, reflecting what Jane Jacobs calls an 'unbuilding' of cities through 'ruthless and oversimplified, pseudo-city planning and pseudo-city design' (1961:408). The area has been the focus of development pressures for over a century resulting in the acquisition, consolidation and reorganisation of land ownerships. The attraction of this South London gateway, as an opportunity to develop businesses in its small market spaces or a chance of financial return from remaking large areas of land, coexist with the threat of having those very opportunities and spaces taken away.

In this chapter I explore the ambitions of the large-scale public-private development partnership to demolish the shopping centre and the market and the needs of traders, residents, students and office workers who used and activated the spaces. Setha Low and Neil Smith (2006) describe common approaches to the appropriation of individual public spaces, through redevelopment, management and regulation; strategies of privatisation that are now prevalent beyond familiar public spaces of global cities such as New York (Sorkin 1992; Kayden 2000) and London (Minton 2006). However, Low (2006:83) states that these strategies employed by private interests in the taking of small-scale public spaces are 'inadequate when dealing with large tracts of land or urban neighbourhoods' like Elephant and Castle. At these larger scales of urban redevelopment 'an expanded set of strategies are employed' (Low 2006:83). This chapter explores a neighbourhood impacted by similar planning mechanisms, legal arrangements and financial incentives used in London to facilitate comprehensive change at such a scale. I aim to expose the dynamic, which Jacobs recognises, between physical spaces that are acquired by law and livelihoods that are ultimately removed (1961:312).

Processes of making

> 'They are developers, they plan on making money out of it. But they also know that if they make it a better place they will make more money out of it too.'
>
> —Jane Goodwin, Chair of Elephant and Castle community group (Interview 2012)

After being replanned several times and dramatically reconfigured in the 1960s following a major transport redevelopment, Elephant and Castle is being masterplanned again. Buildings are being demolished, roads remodelled, leases run down, and communities decanted under the claim that several decades of decline will be turned around through a comprehensive redevelopment (www.elephantandcastle.org.uk 2012). It is the latest attempt to regenerate this neighbourhood and strategic transportation hub. In 2007, Southwark Council ended years of uncertainty and decades of unrealised plans to select the global property and infrastructure developers, Lend Lease, to produce and realise a comprehensive masterplan. The decision to select Lend Lease over competitor Key Property Investments, a team that included the owners of the Elephant and Castle Shopping Centre, was explained by Councillor Nick Stanton, the leader of Southwark Council:

> The basis of our selection was our judgement of the partner's [Lend Lease's] ability to bring finance, resources, experience and capacity to work harmoniously with the council to add value and to deliver the programme.
>
> *(www.london-se1.co.uk 2007)*

From 2007 to 2010, during which the global financial crisis dramatically unfolded, the developer and the council negotiated and signed an agreement to transform the Elephant and Castle area, promising a future with larger parks, safer streets, better shops and a lot more housing. Half a century after Elephant and Castle was previously transformed via a car-oriented approach to modern planning (1961–1974), it was once again reimagined and reconstituted with new strategies to transform the area.

Both public and private goods are being appropriated through the latest redevelopment. The large-scale remodelling of Elephant and Castle has relied on what Low describes as 'legal and economic strategies' that facilitate ways of 'placing public goods in the hands of a private corporation' (Low 2006:83). There have been repeated attempts to regenerate the area, but it has only been through the alignment of political and economic conditions that this has become possible. The employment and threat of legal mechanisms, such as Compulsory Purchase Orders (CPOs), has also facilitated the acquisition of private spaces, businesses and livelihoods. The public bodies of Transport for London and Southwark Council, along with the influential global development partner Lend Lease and the latest shopping centre owners Delancey and APG, negotiate the spaces of Elephant and Castle and in so doing negate marginal market interactions and daily lives.

Incremental development

Structural changes at Elephant and Castle have consistently followed intensifications of the areas' transportation networks. The settlement, known initially as St. Mary's Newington (1222), developed incrementally in the thirteenth century as a crossing of roads in and out of London. Connections between London and towns and cities across South England led to an expanded network of highways that passed through Elephant and Castle reaching the River Thames. From Westminster Bridge (1750), Blackfriars Bridge (1769) to London Bridge (1176) all routes that crossed the Thames passed through what was to become known as Elephant and Castle. The New Kent Road was built in 1751 to connect the South East of England with Westminster Bridge, and it was shortly after that the Elephant and

Castle public house is understood to have opened, over time lending its name to the area. For several hundred years Elephant and Castle grew, with change accelerating as new roads, rail and tram networks were laid out on top of each other. This cumulative growth continued until, in 1897, the first comprehensive plan was published proposing to improve the road layout. By 1912, when a government memorandum records an initial conversation about the potential redevelopment of Elephant and Castle, the area had trams on the streets, trains elevated above, and Tube lines running below. The noise, pollution and potential hazards caused by this congestion of traffic generated public concern for pedestrian safety.

Until the second half of the twentieth century, Elephant and Castle was defined by a fine-grain mix of streets, roads, rows, places, terraces, yards and alleys. These were the public spaces of Elephant and Castle: at the intersections of the streets and roads, people met, congregated and lingered. As Elephant and Castle grew without formally planned open spaces, public activities occurred on the walkways and pavements at the edge of the streets, and in the public houses, taverns and theatres. This was a time, as Richard Sennett claims, where 'The focus of [this] public life was the capital city' (1977:17). But while the streets of Elephant and Castle were not planned, for the 'special purpose of pedestrian strolling as a form of relaxation', public spaces were connected to department stores, coaching inns and theatres that were also 'open to a wide public' (ibid.). On historic street arrangements, as Jacobs reinforces, 'the casual public sidewalk life of cities ties directly into other types of public life' (1961:57). Even in the early twentieth century, the drinking fountain, the cab stand, the middle of streets, the steps up to the Metropolitan Tabernacle and entrances to pubs, theatres and markets composed the public realm of Elephant and Castle. The streets were how people travelled to work, where they queued for theatres and where they expressed themselves in public.

Modern ambitions

The mesh of streets and terraces fell victim to the first comprehensive remaking of Elephant and Castle (1950s–1970s). Since the twentieth century the prevailing approach has been extensive rebuilding. Even before the Second World War, which brought a renewed momentum for planning in the 1950s, the government's ambition was to establish a 'tabula rasa' to build over the network of streets. Comprehensive plans required significant financial resourcing, which, during the middle of the twentieth century, was provided by the London County Council and central government. Although previous plans were interrupted by economic decline in 1933, they were resuscitated after the war to realise a series of government, educational, office and commercial buildings. At the centrepiece of the realised plans was Europe's first indoor shopping centre, a super-block encircled by a system of roads and roundabouts that would replace the intricate street pattern and the Elephant and Castle pub.

The plans were to remake Elephant and Castle as 'the shopping, business and recreation centre for South London' (Londoners get new shopping centre at

Elephant and Castle 1965). However, when the shopping centre opened to reveal that its three levels of retail space were only partially let, the design, planning and construction were quickly and publicly lambasted. Jacobs strongly critiques these types of 'super-block projects' that replaced intricate streets as they reduce the diversity that small urban blocks offer and limit options for pedestrian movement (1961:186). A 'flight from the street', that William Whyte (1980) describes, was facilitated by the self-contained 'mega-structure', an internalised environment 'where the resemblance to fortresses is not accidental' (1980:85–87). Although the mall of the Elephant and Castle Shopping Centre was built on the former alignment of Walworth Road, and was glazed at either end to extend views inside and out, its giant scale, oversimplified form and surrounding moat failed to engage with people accustomed to navigating street-level pavements.

When the congested streets were built over, the proposed plan was to separate the street into efficient roadways, sunken and elevated pedestrian routes, and open public plazas. The streets which had grown and evolved over centuries to accommodate a diversity of daily lives were removed to make way for the first masterplanned redevelopment of Elephant and Castle. The road intersections were replaced with a multiple-lane gyratory of traffic that wrapped around new buildings with huge floorplates, while pedestrians were separated from the intensity of the vehicular traffic by new tunnels below and barriers above. These were public space systems that, as Whyte (1980) claims, seem comprehensive but in terms of a 'completeness not relevant to most pedestrians'. Whyte recognises that in projects like Elephant and Castle, 'the great bulk of traffic is concentrated on the main connectors' and in the other spaces of these efficient systems there is 'nobody' (1980:84). The shopping centre and its network of public spaces, that included the narrow sunken plaza, became what Sennett terms 'dead' public spaces that have 'no diversity of activity'. Sennett describes that such inactive spaces act solely as 'an area to pass through, not to use' (Sennett 1977:12–13). It is in the public space of the moat, after several decades of infrequent activity, that the market emerged to fill a need, again, for the diversity of activity in the public spaces of Elephant and Castle.

As Elephant and Castle has been reimagined, later phases of development have been tasked with solving the problems of previous attempts at comprehensive redevelopment. For example, the gyratory system of roads was built to act as an armature around the future focus of the shopping centre. Ironically, however, as the *Illustrated London News* (1965) reported, one of the main attractions of the shopping centre was its ability to remove shoppers from the 'noisy, dangerous traffic circling one of London's busiest roundabouts'. Similarly, in 1975, a year after the final phase of the masterplan was completed, a milestone that marked the end of decades of redevelopment, the shopping centre was sold and immediately planned for refurbishment. Over the following years many new initiatives attempted to address what were considered problems created by the first comprehensive redevelopment. These included: the top floor of the shopping centre that struggled to attract retailers being converted into offices; lighting and murals painted in the narrow pedestrian tunnels that were considered dangerous and unwelcoming; and the shopping centre painted pink to cover the discolouration of the concrete panels.

The most enduring of these ideas, however, was the Elephant and Castle Market that was established in 1992 by Urban Space Management (USM), the regeneration company who managed the shopping centre for UK Land. The intention of this low-cost market was to bring visible activity out into the open spaces to attract more people into the shopping centre. The market was formed from temporary structures, including forty-five steel frame stalls that were built and dismantled daily, as well as repurposed shipping containers and food trailers. Elephant and Castle Market was not the well-polished, trademarked and architect-designed Container City that USM display on their website (www.urbanspace.com 2014) or the neatly branded Box Park shopping arcade built out of shipping containers in East London's Shoreditch. The rusting containers of Elephant and Castle Market squeezed inflexibly into the plaza. When the market was open, the goods from the containers spilled out into the space, and when it was closed the once modernist plaza resembled an abandoned, windswept yard. But despite its makeshift appearance, and until the residents were removed from the adjacent Heygate Estate, the market was a bustling and frequently visited public space.

This intensity of Elephant and Castle as a transportation hub and as a gateway into London has attracted many people to work, live and seek entertainment in the area. The market was established in the open plaza beside the shopping centre to take advantage of this centrality. Until the masterplan began to be implemented, the space was connected to a maze of pedestrian subway tunnels, filtering and funnelling people into the Shopping Centre from the busy roads above. Stephen James, who worked for the private company that operated the market, claimed that 'the market works because it is very central and as with all markets it needs footfall' (Interview 2012). This desire for proximity to the high footfall of commuters that passed through the space between the subway tunnels and the train station was reflected in the market traders' interest in the area. Barry, a regular trader for four years, insisted that the 'location' of his stall by the subway tunnel entrances was essential for the successful operation of his business (Fieldwork conversation 2012).

The plaza in which the market operated was central to the transport interchange at Elephant and Castle. It had four entrances from the street level, two from the shopping centre and three underground subways that connected the market to 29 bus routes, two Tube lines, the National Rail station and all the areas of Elephant and Castle beyond the busy traffic roundabouts. The market provided low-cost household goods, clothes, perfumes, groceries and lunch for those living nearby, working in and passing through the area. Before the adjacent social housing of the Heygate Estate was closed to facilitate the latest development, including the removal of over 3000 residents, the market was activated and sustained by people who came across from the mass public housing, east of the shopping centre and by those who commuted through the market. Roland, the market manager, described how the market 'used to [get local people] when the Heygate was there, now the

FIGURE 2.2 The food court within the market provided low-cost lunches to students, workers and residents around Elephant and Castle.

(Photo by author)

Heygate is gone, we solely rely on the footfall of the public transport. . .' (Interview 2012). This combination of commuters, workers and residents who visited the market reflects Jacobs' (1961) claim that 'workers and residents together' can 'produce *more* than the sum of [their] two parts'. Rather than just the flow of commuters at the beginning and end of every day or the bustle of office workers in the market's food court during lunch, Jacobs suggests that workers and residents 'must appear at different times' of the day to make open spaces successful (1961:153).

The market became a space for the people who lived, worked and passed through Elephant and Castle. Its home in the moat, around the windowless fortress of the shopping centre, connected directly to the lower-ground floor of the shopping centre. The shopping centre complex was designed by architects Boissevain and Osmond, originally with shop windows illuminating and animating the sunken plaza. After several decades of adaptation, the façade presented a collage of damaged and boarded up shop windows that left the shopping centre visually disconnected from the market. The plaza was designed to include seating, fountains and trees but by the time the market was demolished in 2021 only a few damaged trees remained. The market existed in the moat despite increasingly deteriorated spatial conditions and complexities of ownership and management. It remained profitable for USM and for most of the regular traders, although the market manager stated that 'everyone is struggling at the moment' without the residents from the Heygate Estate (Interview with Roland, market manager, 2012). Undeterred by such marginal conditions, the market had allowed diverse public lives to emerge, creating one of the most active public spaces in Elephant and Castle.

The structure of responsibilities across the moat were complex. Until 2013, when the shopping centre was sold to investors Delancey and APG, the open public space of the market was owned, along with the shopping centre, by Key Property Investments, a joint venture between St. Modwen Properties PLC and Salhia Real Estate Company K.S.C. Key Property Investments proposed to renovate the shopping centre and extend it towards the pedestrian pavements, thus filling in the moat. The plans were opposed by Southwark Council who wanted the covered shopping centre demolished to create an open mall of connective streets—ambitions reflected in Delancey APG's plans. The low-cost market was managed by USM since 1992, on a lease from the shopping centre. USM operated the market and rented out pitches to individual market traders. While the market operated, from 6am to 8pm, the space was maintained by and was the responsibility of USM; after 8pm the responsibility for the moat returned to the shopping centre owners. While the main area of the moat was owned by the shopping centre there were fragments of land around the edge of the building that were owned by Southwark Council, the Highways Agency, Transport for London and Network Rail. There were also public rights of way that ran through the market and shopping centre that were required to be kept open. As Steven James, who oversaw the operation of the market for USM described, the issues of ownership and management were 'very messy because it was never formalised back in the 60s when the place was built' (Interview 2012).

Ambiguities over who owned or managed the public spaces around the shopping centre contrasted with the privately owned public spaces that have been built across London over recent decades (Minton 2006, 2012; GLA 2011b; Carmona and Wunderlich 2012). Many privately owned public spaces have been realised by developers as a requirement of Section 106 planning obligations and have remained privately owned and managed. But the moat at Elephant and Castle was never formally adopted as publicly owned through an oversight by the council. The lack of clarity even confused the council, as demonstrated by the street level vendors who traded from roughly constructed kiosks by the bus stops. Steven James explained that after the market was established, Southwark Council was approached by several individuals who wanted to set up kiosks on the pavements by the busy bus stops (Interview 2012). Permission was granted, plywood structures were built, and businesses opened. Shortly after, it was brought to the council's attention that the land was not council land but instead the property of St. Modwen, the shopping centre owners. The management of the spaces and leases to the street vendors were swiftly transferred to St. Modwen. Steven James concludes: 'And the council went, "Oh really, oh, sorry about that, there you go St. Modwen, have 'em back"' (Interview 2012). The proposed public spaces of parks, streets and market squares that are being realised as part of the masterplanning are expected to remain privately owned and operated.

The spaces at Elephant and Castle were not easily framed through distinctions between public and private. The market contained multiple forms of public life constantly emerging through everyday actions of gathering, meeting, exchanging, eating, playing and travelling. Since the residents of the Heygate Estate were removed to other areas of London and South East England, fewer people used the market for their daily shopping. However, even in the final years that the market operated it was evident that local office workers and students visited the market at lunch; groups of men played cards by the food court; queues formed at the ATM; and, commuters passed through, as they changed from train to bus to Tube, on their way to work. Beyond the rhythms of planning and redevelopment the spaces of the market were contested differentially through the life of the market, the passage of commuters, the gatherings of visitors and through commuters and researchers taking photographs. Conflicts were negotiated every day as people sat in the tunnels asking for money, as crowds of young people queued to enter The Coronet music venue and as families, commuters and students pushed onto the buses above the market. Roland, the market manager, and several market traders described competition between traders for specific market stalls that benefited from higher footfall by the tunnel entrances. However, as the pedestrian tunnels were gradually closed as part of the redevelopment fewer people passed through the moat and the intensity of daily interactions within the space declined. While the market continued to attract new traders—such as Al, an Afghan, who lived in the nearby Aylesbury Estate and who sold low-cost shoes for several months on a stall he bought from a friend—they struggled to establish a regular presence at the market. As the shopping centre redevelopment became more fraught the market further declined and negotiations over the best locations for market stalls became competition for the remaining customers who visited the moat. 'It's quiet, everyone's complaining', Al commented shortly before he closed his stall for good (Fieldwork conversation 2012).

Full demolition

The moat in which the market operated was filled in as part of the regeneration. The low-cost market in the moat was closed and a craft market was established in a new square on the other side of the railway station. When St. Modwen and their investors bought the shopping centre in 2002 they were attempting to leverage their position to become the council's development partner for the whole of Elephant and Castle's redevelopment. In 2004, as the council were developing their masterplan, the shopping centre was proposed for demolition with the replacement being a new street-level retail section, extending towards Walworth Road through the centre of Elephant and Castle. St. Modwen were confident that if they were not selected as the council's development partner then they would profit from having the shopping centre taken from them through compulsory purchase to achieve the comprehensive plan. However, by the time Lend Lease were selected by Southwark Council the economic conditions had shifted and an early opportunity to buy the shopping centre to demolish it was missed. Meanwhile, St. Modwen had to reconsider what to do with their shopping centre asset.

Although they were unsuccessful in partnering with the council, St. Modwen soon recognised that they had maintained a central position in Elephant and Castle's redevelopment. The image of Elephant and Castle was significantly defined by the shopping centre, making the centre's redevelopment key to Southwark Council and Lend Lease's success. St. Modwen therefore began to publicly announce their intention to completely renovate the shopping centre. To make this feasible, the owners claimed, the new building would need to extend out into the moat and an additional 1,000 new residential apartments would be required in a tower above. St. Modwen did not submit for planning approval, however; they began running down the leases inside the shopping centre and their lease with USM for the market. Jonathan Newton of St. Modwen claimed that they were waiting to see what their Section 106 contribution would be; however, Tom Appleby, of Southwark Council, seemed frustrated with the situation, claiming, 'St. Modwen are playing hard to get' (Interviews 2012).

By 2013, as Southwark Council progressed negotiations with Transport for London on a new Tube station adjacent to the shopping centre and the market, the local authority began to put pressure on St. Modwen regarding their intentions. Councillor Fiona Colley announced a 'renewed confidence in the [redevelopment] project' but less satisfaction with the shopping centre plans. She stated that 'the council has now rejected earlier options to retain some parts of the original [shopping centre] building and will now insist on full demolition' (www.southwark.gov.uk 2013). Within three months St. Modwen and Salhia Real Estate Company K.S.C. had sold the shopping centre to Delancey and APG for £80m in cash, over £50m more than they had bought it for in 2002, thus 'crystallising a significant profit for our shareholders' (www.stmodwen.co.uk 2013).

Scales of development

> 'Lend Lease, St. Modwen and the council, they ended up sort of dancing around each other and trying to work out what was going on . . . in the end it became split.'
>
> —Steven James, Urban Space Management (Interview 2012)

Large-scale relations

Strategies of assuming control of land through redevelopment favour large-scale spaces, economics and power. Beyond the sliver of the moat, ambitious developers and international investors, with access to vast capital resources, jostled for control. Less visible and obscured by the prominence of the latest large-scale redevelopment exist failed partnerships with the local community, small businesses wound down and individual concerns overlooked. The agreement between Southwark Council and the Australia-based developers, Lend Lease Corporation Ltd, was formed around Southwark Council's position as the majority landowner, who until August 2013 maintained control of the Heygate Estate, and Lend Lease's track record of large building and global redevelopment projects. The latter operates in over fifty countries focusing on property and infrastructure. In 2019 Lend Lease reported a 'development pipeline' of almost $100 billion and $467 of profits after tax (www.lendlease.com 2019). The *New Statesman* claimed in 2013, using information from local campaigns against the Elephant and Castle development, that Lend Lease were due to make a profit of £194 million on their redevelopment. The London Borough of Southwark is one of thirty-two London Councils and spans from the River Thames to Dulwich and Borough to Rotherhithe. Southwark owned the site of the Heygate Estate that they exchanged with Lend Lease for £55 million; as part of the agreement, and in preparation for handing over the land, the council evicted the residents and cleared the site, at an estimated cost of £65 million.

Unlike the previous major reconstruction of Elephant and Castle, which was finally completed with the opening of the Heygate Estate in 1974, local authorities in London have not been able to implement major redevelopment on their own. Poor in capital but rich in land, councils like Southwark have come to rely on partnerships with well-funded developers and teams of private consultants. Long-term public-private relationships have led the regeneration at Elephant and Castle, reflecting what Rob Imrie and Mike Raco describe as central government's continued 'dismantling of the postwar welfare state' (2003:12). Rather than reversing the market-based government policies that began under the Thatcher administration, Imrie and Raco describe the New Labour government's increasing reliance on partnerships for urban redevelopment, leaving the state to take on a 'facilitating role' (2003:12). Now that local government can no longer engage on its own at this scale of the city, large developers with private capital investors are fundamental to the process. But this coming together of public agencies and private corporations creates ambiguous relationships and a development process that the council does not always control. As Low claims, planning mechanisms are employed that 'utilize normative governmental procedures but are manipulated for private ends' with the long-term consequences being an 'impoverishment of the public realm and limited access to public resources' (2006:83–84).

The public-private partnership operated at a neighbourhood, rather than a building scale. A large-scale development was necessary for Southwark Council to sufficiently profit from the value of the land, which it owned, and for Lend Lease to commit sufficient resources and ensure a financial return. As part of the partnership agreement between Lend Lease and Southwark Council, the developer was required to provide a social infrastructure of parks, open spaces, recreation and transportation enhancements. Section 106 of the 1990 Town and Country Planning Act, which states obligations to be placed on developers to offset the impact of their developments, has become an essential tool for local authorities with limited resources of their own. Through Section 106 agreements local authorities can provide much needed improvements to their boroughs—or comprehensive transformations as in the case of Elephant and Castle—while simultaneously relinquishing their previous responsibilities for the long-term management, security and maintenance of public spaces. Since the post-war period, when Elephant and Castle was rebuilt by the state, with funding and influence from the London County Council (LCC) and central government, approaches to urban redevelopment have changed. However, then as now, and throughout the shifting roles of government, free markets, communities, private-public partnerships and local initiatives, the conditions for achieving a successful redevelopment have remained the same: to engage with large-scale masterplanned redevelopment public agencies and private businesses need to be able to secure a sufficient scale of finance and land resources.

There have previously been several attempts to regenerate Elephant and Castle at a small scale. After the refurbishment of the shopping centre by UK Land, providing for the establishment of the market, two notable and public-led small regeneration initiatives began in the 1990s: *Elephant Links* and *Elephant Impacts*. These community enterprises included attempts to address physical conditions through tree planting, lighting, painting and the design of small structures. While these initiatives unfolded, and the appearance of Elephant and Castle was being aesthetically transformed, a survey of the Heygate Estate by Southwark Council in 1998 indicated the potential of refurbishment rather than demolition. The survey found that the architectural structure of the Heygate Estate was sound and that renovating the buildings was the best option. However, as local resident and council employee Paul noted, the recommendations would have only facilitated a local scale of regeneration, which would fail to 'release any of the value of the land' (Interview 2012). Refurbishment would not have facilitated building new market-rate residential units, and as a result it would not have enabled large Section 106 contributions. In addition to releasing the value of the land, it was claimed by opponents to the regeneration that Southwark also aimed to decrease the proportion of social housing in the area, which was previously at 75%, and increase the percentage of market rate housing to 75% (www.urban75.org 2013). As Paul described:

> I don't know how true . . . there was a kind of suggestion that . . . Elephant and Castle had at the time about 75% social housing, which was too much, way too concentrated, and it needed to be turned into a more mixed community . . . and magically everyone's lives would improve.
>
> *(Interview 2012)*

In the 1990s, when the idea of remodelling Elephant and Castle for the second time began to be seriously discussed, there were several community groups

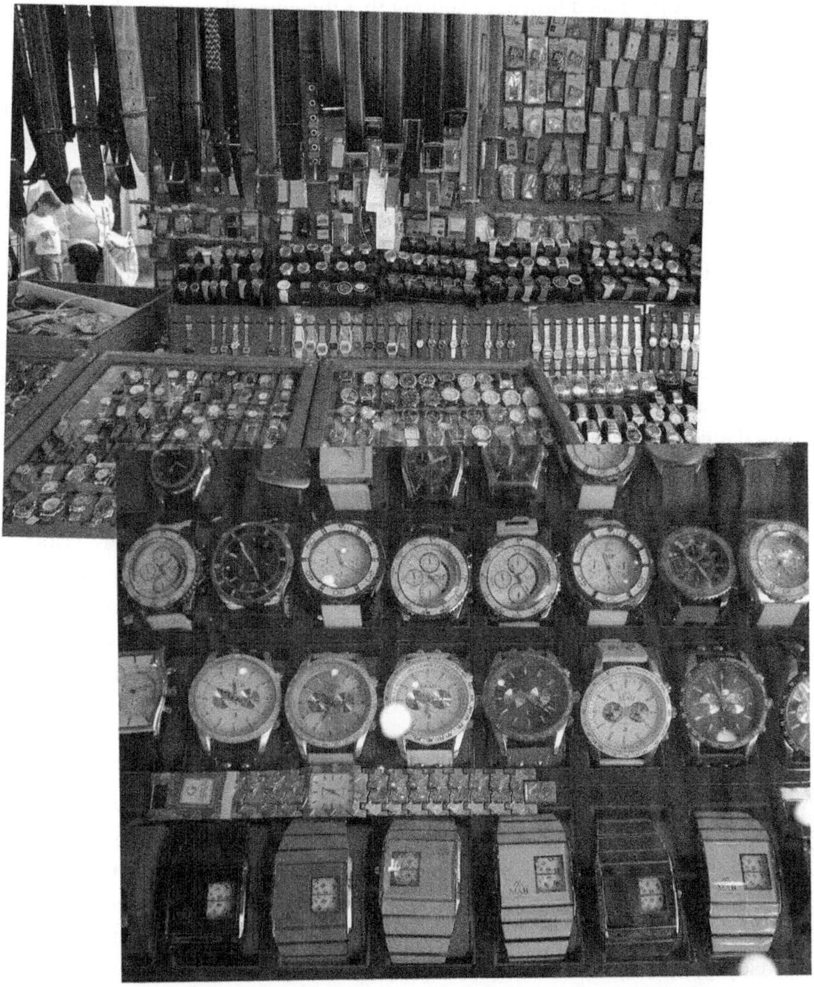

FIGURE 2.3 Traders selling fruit and vegetables, clothes, shoes, jewellery, suitcases, carpets and perfume—alongside people distributing religious books—were integral to the public life of Elephant and Castle.

(Photo by author)

involved. The Southwark Land Regeneration Partnership (SLR) was a private sector initiative that engaged in a three-way negotiation with local tenants and the council. Although the current development partnership, between Southwark Council and Lend Lease, claimed that the SLR 'helped clarify its aims for the area and led directly to the establishment of a strategic vision that underpins the current programme' (www.elephantandcastle.org.uk 2012) negotiations with the local community collapsed. The community website, *Southwark Notes* provides an explanation of the failed process, claiming that the council and the SLR developers 'couldn't agree on who would get what share of the profits' (www.southwarknotes.wordpress.com 2012). Paul offered an alternative perspective, in which the regeneration 'bit by bit . . . fell to pieces' with the rhetoric from the council being that 'community groups could not work effectively together' (Interview 2012).

Since Southwark Council began to look towards comprehensive redevelopment, claims of conflicts of interest between the different roles of the council—as landowner, developer and planning authority—began to be voiced. The regeneration of Elephant and Castle has been guided by the Elephant and Castle Supplementary Planning Document (SPD) (Southwark Council 2012), which was drafted by the council as the planning authority. Several months after the SPD was adopted, Southwark Council and Lend Lease as development partners submitted a planning application to progress a major phase of the redevelopment. Southwark Council received and approved this application as the planning authority. Paul considers that these multiple roles leave the council with 'a kind of split personality between a developer partner and a planning authority' (Interview with Paul, a local resident, 2012). Ambiguities in the role of the council have led many critics to question the objectives of the local government and their relationships with others in the development. These concerns have been exacerbated as many employees and councillors from Southwark have moved across to work for Lend Lease and other developers through the masterplanning process (www.35percent.org 2021).

Comprehensive redevelopment?

Although the two major stages of rebuilding Elephant and Castle (1950s–1970s and since 2004) have appeared comprehensive, their realisation has progressed sporadically and in piecemeal ways. Despite the bold masterplans and strong rhetoric, the large stakeholders of the redevelopments have also struggled to fulfil their ambitions. In 1947, the *London Evening Standard* reported on the first Elephant and Castle redevelopment as a plan that was to take between five and fifteen years to complete. However, despite the clarity of the plans, alternative visions continued to be drawn up over the next decade, even after the construction of the roads and roundabouts had got under way. Rather than one comprehensive plan that would be built in a few years, this refashioning of Elephant and Castle was a series of plans, buildings and infrastructures orchestrated by competing agencies and ambitions that would take over two decades to complete. As landowners and economic conditions have changed, masterplans by Foster and Partners, Martha Schwartz and Partners, SOM, Allies and Morrison and Make Architects have been authored—and many abandoned. This piecemeal progress is reflected in the redevelopment of Elephant and Castle's public spaces, with Tom Appleby from Southwark Council, describing the public realm strategy as 'just kind of incremental at the moment' (Interview 2012).

Even since 2002, when the London Plan identified Elephant and Castle as a key area for growth, many alternate plans have been drawn up. Local resident and council worker Paul described 2004 as the 'zenith of this utopian narrative' for remaking Elephant and Castle as 'there was plenty of money knocking about' (Interview 2012). Over the next few years, a comprehensive plan was promoted that included the removal of the shopping centre, however shortly after Lend Lease was selected as development partner, the global financial crisis unfolded and the conditions for development changed. From 2008, the pressure on Southwark Council intensified as they had already begun to empty the Heygate Estate of its residents. Southwark Council had expended significant sums on private design and planning consultants and Lend Lease, their chosen development partner, was struggling to finance their projects due to the economic downturn (www.building.co.uk 2008). In contrast to the first replanning of Elephant and Castle, which was defeated by the financial collapse of the 1930s, this redevelopment, or at least something like, had to be built out. However, the council was not in 'a particularly strong bargaining position', Paul notes, 'if the deal with Lend Lease had failed then not only would it have been a big disaster [politically], but financially it would have been a big disaster' (Interview 2012). So as a result, plans were redrawn, contracts were renegotiated and the phases of redevelopment unfolded in a piece-meal and less coordinated manner. Gavin Poynter describes a feature of a market-led planning system that requires the 'dilution of the role of the government' (2009:134). At Elephant and Castle, as the global financial landscape collapsed, the local government of Southwark Council had no option but to further commit to the uneven relationship that they had already contracted with Lend Lease.

Concealed conversations

As the relationships between the public-private partners became less balanced, details of the agreement were increasingly concealed. Reflecting Low's assertion that such relationships between public and private organisations are 'harder to identify' (2006:83), local residents and community blogs began to criticise the opaque nature of the negotiations (www.southwarknotes.wordpress.com 2013). Poynter explains that in such situations the 'removal of public scrutiny over the financing arrangements' can be deemed necessary to move developments forward (2009:134). The opaque nature of negotiations and agreements has remained one of the greatest concerns of local community organisations and critics of the redevelopment. Steven James, from USM stated: 'once the main developer has been chosen it has gone a lot more behind closed doors, at least with the decision making part of it' (Interview 2012). Attempts at privacy, for reasons of 'commercial sensitivity' were undermined in 2013 when Southwark Council released a poorly redacted version of the regeneration agreement with Lend Lease. Details of the regeneration project, which were until then kept private between the two partners, became publicly accessible courtesy of the local blog, *Southwark Notes* (www.southwarknotes.wordpress.com 2015). Financial details of the agreement revealed the significant profit for Lend Lease alongside the potential losses to be incurred by Southwark Council.

The piecemeal and discontinuous approach of the large-scale redevelopments which have transformed Elephant and Castle have also led to uncertainty and disadvantage. As large-scale plans take decades to be complete or are shelved due to less favourable economics, smaller-scale businesses and local residents find it difficult to consider their own futures. In 1933, a local businessman wrote to the Ministry of Transport protesting the lack of information about the impending 'improvements' and the impact this was having on their businesses. In March of 1933 the London County Council wrote on their behalf, that 'some of the tradesmen in this area are in a somewhat embarrassing situation owing to the uncertainty of the improvement' (Letter from National Archive 1933). In the same year Edward Strauss of Borough High Street wrote to the ministry: 'The present position inflicts great hardship on business people in the area affected by the scheme because their power to deal with their own property is severely circumscribed' (Letter from National Archive 1933). Although the plan for improvements, which included the demolition of many businesses, houses and streets, were abandoned until after the war, the uncertainty greatly impacted confidence to plan and the ability to do business in Elephant and Castle.

This unsettlement corresponds to conflicts between local authorities, developers, market traders, medium-sized businesses and residents in Elephant and Castle today—from the contrasting scales of the super-block shopping centre to the individual market traders and shoppers. The shopping centre was, for several years, under threat of compulsory purchase for Southwark Council and their development partners to realise the potential of the central site. As a result, the many leaseholders chose not to renovate but rather to rely on low-cost initiatives and maintenance. Steven James, from USM, described their frustration: 'We would absolutely be willing to invest in really trying to build the market up but you have got to have some kind of certainty' (Interview 2012). He went on to explain the difficulties that traders faced who 'don't want to invest in a brand-new stall with all the trimmings if they are going to be told they have to get off next week.' James described that USM felt 'left in the dark' and that for almost ten years he had waited for assurances:

> I have sat on various committees, in all their different guises at Southwark . . . and no-one can ever tell really what is going to go on. There were tenants' committees at that place, meeting every six weeks for about five years and they got no-where because in the end it will be decided by the developer and Southwark.
>
> *(Interview 2012)*

This uncertainty was compounded by the lack of investment in the increasingly deteriorated spaces of the market and shopping centre, an appearance that furthered the arguments for replanning set out by the development partners. As Ben Campkin describes of Elephant and Castle: 'demolition discourses have long contributed to its [the shopping centre's] blight' (2013:71).

The large-scale, uncertain and piecemeal approach has tended to overshadow spaces, activities and people: the space of the moat, the activity of the market and the people whose lives were entwined in these spaces were consistently marginalised by the redevelopment. For £27 anyone could turn up at Elephant and Castle Market and rent a stall for the day, as a casual trader (Interview 2012). There were some basic rules of trading that were set out in the market handbook—but if there was a stall available in the morning then anyone could rent one and establish a presence in the public spaces of London. Madanipour claims, however, that these 'marginal public spaces . . . are not on the list of priorities' when local authorities are considering redevelopment (2010:113). The market, and even the shopping centre, did not contribute to an 'image and marketability' raising questions as to whether the moat, its market and the people who worked in and used the area had a future in the space at all.

USM's Steven James recognised that the uncertainty over the large-scale redevelopment impacted the market traders the most (Interview 2012). Barry, a regular trader at the market, stated that his immediate concern was how quiet the market had become since the Heygate Estate had been decanted, but he was not 'making any plans for the future—until the regeneration [was] known' (Fieldwork conversation 2012). However, while the residents of the Heygate Estate were offered a 'Right to Return' to the area following redevelopment (www.southwark.gov.uk 2016)—a process that was not without criticism for its lack of fairness—the traders, some of whom have been trading from the market for almost fifteen years, were not given any assurances that they could remain. Both the Market Charter and the SPD described the need for a market at Elephant and Castle, but Tom Appleby, the Project Director for Southwark, questioned whether the existing traders were 'the right operators to go on that new space' (Interview 2012).

Taking through remaking

> 'It [the Elephant and Castle Market] could be improved from what it is today, but, when you are under constant threat of having it taken away or bulldozed and you don't know what the future is. . .'
>
> —Steven, local manager (Interview 2012)

The remaking of Elephant and Castle has facilitated processes of taking. Investors with money have strategically bought up property while organisations with power have run-down (or taken back) leases, cleared estates, and sought out CPOs for themselves and on behalf of others. As occurred in its first reorganisation, the scale of ambition for Elephant and Castle has overshadowed the spaces, activities and certainty of small businesses, property owners and residents. As Fiona Colley, local councillor for the area wrote in *The Guardian*, 'Elephant and Castle regeneration will transform the area, not just one [Heygate] estate'. Colley, who threatened to use a CPO to force the previous shopping centre owners, St.

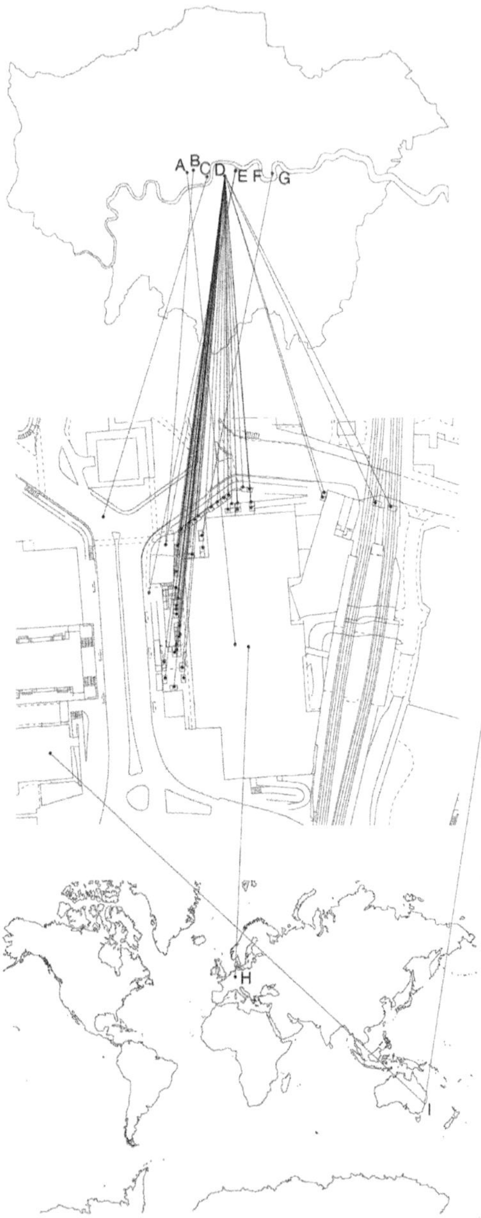

FIGURE 2.4 Decisions about large-scale redevelopment were made in planning departments and corporate offices far from Elephant and Castle, while the daily remaking of the market was decided by managers and traders who worked in the market.

(*A:* Delancey B: Transport for London; C: Department for Transport; D: Vendors; E: Southwark Council; F: Greater London Authority; G: Urban Space Management; H: APG; I: Lend Lease) (Diagram by author)

Modwen, to replace rather than renovate the existing building even extended her ambitions beyond the Opportunity Area of Elephant and Castle: 'our vision is to transform the whole borough' (www.theguardian.co.uk 2013).

Rationales for appropriation

Transportation proved a useful rationale for appropriating this part of London. As new roads and train lines have been built through the area, Elephant and Castle has been reconfigured. Increased movement into London from the south has consistently required streets to be widened and others to be built, resulting in the demolition of buildings and the displacement of businesses and residences. In 1876 St Mary's Newington church was demolished by the Metropolitan Board of Works to accommodate the widening of Newington Butts while in 1892 talks began about the demolition of the Elephant and Castle public house after which the area had gained its name. Compensation for the appropriation of these individual buildings was achieved through the rebuilding of the pub and the building of a new parish church on Kennington Park Road. However, by the beginning of the twentieth century, cars were vying for space on the streets. This new motorised vehicular traffic, that Jacobs (1961) describes as an 'instrument of city destruction' where 'city streets are broken down into loose sprawls' (1961:338) became a modern pretext for redeveloping the area.

Despite her criticism of highway developments, Jacobs warns against demonising the car, asking 'how much of the destruction wrought by automobiles on cities is really a response to transportation' and how much is 'owing to sheer disrespect for other city needs, uses and functions' (1961:339). From 1912, for almost half a century, traffic solutions were discussed, dreamed and drawn up that included comprehensive development of much wider areas of Elephant and Castle. In 1930 the Charity Organisation Society wrote to the Ministry of Transport (MoT) to complain that 'the dwellings of 2000 Citizens of this Borough are to be swept away' to 'carry out a grandiose scheme for relieving the traffic difficulties at Elephant and Castle' (Letter from National Archive 1930). Rather than appropriating individual buildings as had occurred in the past, the national and city governments became increasingly intent on using legal mechanisms, or creating them if they did not exist, to reorganise the entire area. In 1945 a meeting between the LCC, the MoT and the Ministry of Town and Country Planning (MoTCP) the Town and Country Planning Act was discussed necessary to 'acquire, not scattered properties, but the whole of the area' to guarantee the Elephant and Castle development. The potential benefit from 'rate income' and 'income from rent' was presented as a further rationale for appropriating a larger area of Elephant and Castle, even including 'some of the surrounding properties so that the council might recoup itself by granting commercial leases' (Meeting notes from National Archive 1945).

The bomb damage caused to buildings across London during the Second World War provided another argument for appropriating land through comprehensive planning. With the worsening of the economy in 1933 the redevelopment of Elephant and Castle was cancelled due to the high cost of buying up the land. However, it was anticipated that after the war ended ambitions for a comprehensive remodelling of Elephant and Castle would be realised. This post-war optimism proved difficult for the LCC to capture, leading to a meeting with the MoTCP in 1945, where the Ministry 'felt that the Council's proposal was rather like the man who wants to lick jam off the stale crust instead of eating the lot' (Meeting notes from National Archive 1945). Although the LCC was considered too cautious in the scope of their proposals, a year later they stated that 'bomb damage has eased the difficulties of acquisition'. Acquiring tracts of land, the MoTCP encouraged, 'would be very profitable to the council and that in the interests of the neighbourhood the council should acquire a much larger area' (Meeting notes from National Archive, 1946). In the years to follow schools, alms-houses, streets of houses, in addition to the Elephant and Castle Hotel and Pub and the Trocadero Theatre, were taken and demolished.

In the latest replanning, negative narratives have continued to be employed as a rationale for development. Paul, a resident and council employee, described how 'the council did its fair share of talking down Elephant and Castle to kind of stimulate the idea of an obsolete place, a failing place, a place that is full of crime, and noisy, and dirty' providing a pretext for the partnership of Southwark and Lend Lease to claim that they 'can change all this' (Interview 2012). Negative representations facilitated by the council continued to influence how Elephant and Castle was perceived, with newspaper articles informed by negative visions framed by the council then quoted back by politicians. For example, Charlotte Philby of *The Independent* describes the Heygate Estate as 'a deserted scene from an apocalyptic movie' (www.independent.co.uk 2010). In the article titled 'Living in Ghostland—the Last Heygate Residents', Philby fails to differentiate between the conditions of the estate emptied of residents and what it was like before the regeneration was imminent. Although she describes residents who for a long time found it a 'happy, safe place to live' she does not recognise that it was the talking down of Elephant and Castle and the uncertain prospect of eviction that led many residents to leave. A few years later Fiona Colley, a politician writing for the *Guardian* newspaper in support of the redevelopment, substantiated her praise for the Heygate regeneration by quoting Philby's article: 'Commentators have called it a "crime-racked labyrinth of grey high-rise blocks"' (www.theguardian.com 2013).

These negative representations of a deteriorated Elephant and Castle have been frequently cited to vindicate decisions made. While the deals between the council and its development partner concern residents and traders, as Jacobs describes of her experience in New York, 'it has all been decided before they [the community] are heard' (1961:406). Jacobs describes the workings of public planning meetings, and the helplessness and futility that is felt by local people. At Elephant and Castle, insecurities were exasperated by suspicions that decisions were first reached in private at a senior level between developers and the council.

Jane Goodwin, the chair of a developer-funded community group recognised this experience: 'One of the criticisms will always be, that we have come up with all these suggestions but how do we know that you are actually taking them on board. Well, all of them will not materialise, but some will' (Interview 2012).

By prioritising interests of its development partner, the council failed to represent residents and other businesses. In the first post-war redevelopment, the state appropriated swathes of the area, and since 2004 they have facilitated another taking of Elephant and Castle. The council's partnership with Lend Lease and their transfer of assets was enabled at the scale of the Opportunity Area. Successive national governments have left local authorities in the UK impotent to develop their boroughs without private partners, both legitimising these public-private relations and increasingly necessitating them. The first demolition of Elephant and Castle began shortly after the passing of the Town and Country Planning Act of 1947, allowing local authorities to appropriate private property for redevelopment. The Planning and Compulsory Purchase Act 2004 offers a further tool that modifies the definition of public benefit to favour compulsory purchase to facilitate large-scale economic redevelopments. It is through these mechanisms that the Heygate Estate was emptied of its residents and handed over to Lend Lease. It was with the assistance of this Act that the demolition of the shopping centre and the market has unfolded.

Describing the process at Elephant and Castle, Loretta Lees claims: 'Regeneration in London has become little more than the private sector building expensive properties' (www.theguardian.co.uk 2013). Development has been transformed into a transaction where the state provides a range of mechanisms for private interests to appropriate buildings and spaces. In doing so, Southwark has developed an economic imperative but lost sight of other responsibilities. Local resident and council worker Paul concluded:

> I think that if the council had taken the position of putting across to Lend Lease they had a democratic obligation to deliver these sort of things across the borough: you have your bottom line and we have our bottom line as well, and our bottom line is a different type of bottom line to your bottom line, it's a kind of social bottom line and sustainability bottom line. But I am not sure how much those arguments were really used and we can't tell because it wasn't transparent and if they were transparent I think that a better deal could be achieved.
>
> *(Interview 2012)*

Inevitable displacement

Displacement of homes, businesses and public activities has been one of the results of this large-scale rebuilding. A vague rhetoric about providing support for the relocation and accommodation of existing businesses was offered by the previous landowners and developers. Ian Fraser of the Traders' Association explained the situation for local businesses:

> The problem is that history isn't on our side. There is no other scheme of this size and complexity of regeneration where the independent retailers have survived.
>
> *(www.london-se1.co.uk 2007)*

When asked whether existing businesses would remain part of a renovated shopping centre, Jonathan Newton of St. Modwen clarified: 'I would be very surprised if any of them stayed, to be honest, just because I don't know how many businesses could wait three years from moving out, to start their business again' (Interview 2012). Newton described that the owners had not begun their redevelopment, but for several years they had only been offering short-term leases to prepare for clearing out the centre. He claimed that he was keen to help out the existing businesses, but clarified: 'Don't get me wrong, the rents will increase'.

The displacement of the market and the removal of its traders has created, as Don Mitchell and Lynn Staeheli describe, 'the city in an image attractive to tourists, middle and upper-class residents, and suburbanites' (2006:144). Mitchell and Staeheli's description reflects individual accounts that the council was attempting to increase the number of home-owners in the area and the suggestion from Lend Lease that they were keen to increase the presence of global brands (New London Architecture 2012). Steven James, USM market manager, explained that neither Southwark, Lend Lease nor St. Modwen had an obligation to keep the market or its traders because it was a private market (Interview 2012). He believed that markets reflect the community that they are in, and while he was keen that USM remain involved, he was personally more concerned for the traders themselves:

> Ultimately, we want to see that the traders are OK, because for us it would not be the loss of our jobs and our careers and our companies if we weren't involved. For hundreds of people who have worked down there or who have lived with people who have worked down there, it would be.
>
> *(Interview 2012)*

Markets are considered to offer significant contributions to urban development efforts. The SPD for Elephant and Castle's regeneration (Southwark Council 2012), the document from which the redevelopment was led, praises the role of markets in enlivening town centres and having economic and social benefits. More specifically, it states that the plans for Elephant and Castle that include a new market space will 'give access to fresh fruit and vegetables, supporting local producers' (Southwark Council 2012). The former market, however, only had two fresh fruit and vegetable stalls and had no stands that represented 'local' produce. The new market does not provide space for traders selling low-cost household goods, clothes and food. The website for the new Elephant Social Market highlighted the plans for the new market square:

> Expect a delicious line up of Street Food, farmers market traders, cooking demos, local designers & crafts, live music, art and workshops. Working with local charities and community groups, Elephant Social Market will be a vibrant hub for the local neighbourhood celebrating art, food, music and conversation.
>
> *(www.theartworks.london 2017)*

The redevelopment at Elephant and Castle is considered less committed in improving the lives and businesses of local people. Instead, the council and their development partner adopted strategies to either dilute the presence of the existing communities by attracting new people to an Elephant and Castle reimaged, or to deny their existence by removing the spaces in which they found presence (www.southwarknotes.wordpress.com 2014). Mitchell (1997) describes redevelopment processes and changes in legal structures that specifically target homeless people and he challenges the implication on public spaces and on those who use them: 'these laws attempt not just the annihilation of space, but also the annihilation of people who live in it' (Mitchell 1997:305). While homeless people were less evident at Elephant and Castle, it was in the small and interstitial spaces of the moat, the railway arches and the pavements, that low-cost businesses flourished, and it was through their removal—aided by a range of legal and planning mechanisms—that these unique livelihoods were extinguished.

Negotiating public spaces

The public spaces that support everyday lives in Elephant and Castle are expendable and negotiable. Low and Smith attempt to differentiate public space from private space through 'rules of access, the source and nature of control over entry to a space, individual and collective behaviour . . . and rules of use' (2006:3). The SPD for Elephant and Castle reflects this ambition to make 'clear the distinction between public and private space' (Southwark Council 2012), but how this will be achieved remains unclear. The blurring of public and private space is recognised by Akkar Ercan as a threat to the public space of post-industrial cities. He describes the 'different shades' of publicness that now exist and challenges planners and developers to resolve the ambiguities that arise through redevelopment (2010:48). The open spaces planned as part of the Heygate Estate's redevelopment were intended to be publicly accessible. But it is expected that the spaces will be controlled and managed by the developer, Lend Lease, rather than being formally adopted by Southwark Council. Tom Appleby stated that: 'The council would want full public access and rights of way over it' but concedes that 'it is a cost thing' and that 'Lend Lease will want to maintain it to a level, or standard, that we [the council] might find more difficult to achieve' (Interview 2012). In contrast to the open space of the moat that was in the ownership of the shopping centre and was loosely managed by both Delancey APG and USM, there are increasingly competing interests to control the new public open spaces in London. As local authorities withdraw from their obligation to build and maintain open public spaces, private developers, consultants and landowners welcome opportunities to step in. Conclusions to *Open Space; an asset without a champion*, a survey of real-estate advisors, developers, investors and public entities by The Urban Land Institute (ULI) and Gensler, include:

> It is within the remit of key figures within the real estate industry to champion the implementation of existing collaborative models and the development of new ones.
>
> *(2011:20)*

The private developer provision of accessible open space, which for the purposes of planning applications is frequently described as public, can meet many developer ambitions: first, the developer can fulfil their Section 106 obligations through the provision of publicly accessible space; second, these spaces can provide an attractive landscape setting to enhance the image and value of the wider development; and third, by building these spaces the developer is well placed to provide the management resulting in long-term control over what the area looks like as well as who can and cannot use the new 'public' spaces of the city.

The commercial sector is well positioned at Elephant and Castle to benefit from the new public spaces. In the ULI/Gensler report, all those surveyed recognised the 'commercial value of open space' (2011:7). In authoring the report, the ULI cite the principles of urbanists Jan Gehl, William Whyte and Jane Jacobs. However, they fail to recognise Whyte's concerns that 'many businessmen have an almost obsessive fear that if a place is attractive to people it might be attractive to undesirable people' (1980:60). In many North American cities, as examined by Whyte, it is the homeless that are the 'undesirables'. Mitchell and Staeheli recognise that 'considerable effort has been expended in figuring out how to regulate the homeless from public property, or how to expel them altogether' (2006:151). In Elephant and Castle there were occasionally drunk and homeless looking men in the shopping centre and people asking for money in the narrow pedestrian subways, but they may not be the only people managed-out of the new public realm. As the new privately managed public spaces are built, following the demolition of the subways and the shopping centre, small low-cost businesses, market traders, groups of men playing cards and teenagers hanging out will become less acceptable. Furthermore, Whyte describes, 'many corporation executives who make key decisions about the city have surprisingly little acquaintance with the life of its streets and open spaces' (1980:60). This observation resonates with the situation at Elephant and Castle where the public life of its market and the shopping centre and in the low-cost retail and community spaces have been extinguished through redevelopment led by corporations based far away from where the decisions will be felt.

Conclusions

> 'It's had the terrible misfortune to be regenerated twice: once, in the 1960s, to convert its poor residents into car-driving modern citizens; and again today, to re-programme them into the kind of latte-supping pedestrians you see on architects' drawings.'
>
> —Tom Dyckhoff (www.guardian.co.uk 2012)

I draw four key conclusions from these accounts of Elephant and Castle. First, I identify that the repeated redesign and reconstruction of the area has facilitated large organisations and corporations to appropriate land and displace existing activities. Although I describe smaller scale contestations between market traders and the local authority and conflicts between young people and the police, the remodelling of transport infrastructures provided a firm narrative to facilitate the demolition of buildings and public spaces—with masterplanned change supporting the transfer of large swathes of land between different owners and organisations. The development partners found planning regulations and legal mechanisms were useful to appropriate properties from owners less willing to sell. Section 106 of the Town and Country Planning Act 1990 and the Planning and Compulsory Purchase Act 2004 were employed by Southwark Council to facilitate the development. This involved forcibly displacing tenants from the Heygate Estate so that a cleared site could be handed over to Lend Lease. Southwark Council also threatened the use of CPOs in order to ease the eviction of remaining businesses and ensure demolition of the shopping centre. The local authority dismissed the potential of smaller-scale initiatives to improve Elephant and Castle while also rejecting the opportunity to renovate older buildings, resulting in residents evicted, businesses closed and public spaces demolished.

Second, I have highlighted that claims of comprehensive masterplanning unfold in a piecemeal and incremental manner. To realise the *Elephant and Castle Regeneration* the masterplan needed to be regularly amended, and agreements between Southwark and Lend Lease had to be renegotiated. Planning law approved by the UK government facilitated the development, a process that was further enabled by local and metropolitan governments as planning controllers. The lack of transparency or public scrutiny of the agreements formed between the development partners, in particular when contracts were renegotiated following the financial crisis, made the process more contested. The power that the private developers gained contrasted with Southwark Council's position who were left with few options in moving the redevelopment forward. Having already displaced thousands of residents from their homes and unable to financially control the process themselves, the council kept providing the developers with favourable terms. As local resident, Paul, described,

> One of the main problems with Elephant and Castle is that as a regeneration process it has been going on so long. It is now so politicised, that every

> administration needs to be the one that on their watch they get a spade in the ground.
>
> *(Interview 2012)*

Third, I recognise that the regeneration of Elephant and Castle has proceeded with significant uncertainty for the local authority, private, developers and market traders. The faltering progress of the regeneration was impacted by changes in UK and local governments, a global financial crisis and the varying abilities of the development partners to finance the construction. As Councillor Nick Stanton, leader of Southwark Council in 2007, described:

> We're very conscious that we've been talking about the need to regenerate the Elephant and Castle for a long time, and it's paramount that we get on and show people that we're serious about actually delivering.
>
> *(www.london-se1.co.uk 2007)*

The uncertainty for businesses, residents and landowners has been intensified by Southwark Council who threatened to compulsorily purchase many homes and buildings. Furthermore, businesses were hesitant to stay in the shopping centre as the long-leases they had were run down by St. Modwen so that the owners could redevelop if their proposals were approved. The conditions of the area have been derided by the council, the developers and the press: the visible deterioration of the area and its lack of 'national' retail outlets provided reasons for Southwark Council to 'talk down' Elephant and Castle, while Lend Lease's Development Director Rob Deck joked that he promised Peter John, the leader of Southwark Council, a Starbucks coffee outlet in the new development (New London Architecture 2012). Despite St. Modwen's claims, while they owned the shopping centre, that they 'do a lot of stuff with the community', they also recognised that none of the existing businesses would remain (Interview with Jonathan Newton, shopping centre manager, 2012).

Finally, I conclude that during masterplanning individuals and small-scale businesses with less power than the council, landowners, developers and managers have been rendered invisible. If a hierarchy of Elephant and Castle Market can be imagined, large organisations such as Southwark Council and Lend Lease would invariably be near the top: together they direct the regeneration and they approve or decline planning decisions. Below the development partners would be the shopping centre owners who are key stakeholders in the regeneration, and then we would find the public space managers, such as USM who initiated and operated the market. Individual traders, their customers, Roland, the market manager, and the market cleaners appear to have had less voice in the regeneration and know little of the decisions that were made. Steven James of USM recognised that without the traders the market would not be there, but he was also aware that if the market were removed, 'it would not be the loss of our [USM's] jobs and our careers and our companies' (Interview 2012). Despite their important role in maintaining and activating the public realm, I have found that the traders, cleaners and managers employed by the market were some of the least considered individuals during the development process.

This chapter has described how the large-scale regeneration of London's Elephant and Castle neighbourhood has been contested between the local authority, developers, residents and businesses. Negotiations over land for development, commercial activities and everyday activities reveal contrasting opportunities to engage in occupying, using and transforming public spaces. Despite the liveliness of everyday spaces like the market, comprehensive visions of the area take precedence over lived experiences. By removing the public space of the moat and the market, many traders have lost their businesses and livelihoods. The lives of residents and small business owners, who defined the distinctive market space of Elephant and Castle for many years, were unable to be heard beneath the global financial ambitions that drove the redevelopment. Steven James of USM explained, 'It is a general market, so it relies on the people, a general market for local people' (Interview 2012). As the local people have been relocated and the redevelopment is built out, plans for the area will produce different public spaces, attracting different people. As James described: 'Markets have to reflect the people that are using them, and that is what the Elephant does'.

Note

Fieldwork and discussions from this research into Elephant and Castle market have also been published in: Wall, E. 2021. Working with uncertainties: Living with masterplanning at Elephant and Castle. In: T. Warterman, J. Wolff, and E. Wall (eds.) *Landscape citizenships*. Oxon: Routledge.

3
PLACE AS PROPERTY

> 'I think in terms of a finished product . . . you can only achieve a degree of finish in areas such as this on the basis of land ownership'.
>
> —Julian Dean, planner for the City of Westminster (Interview 2012)

For three decades, the West London canal from Little Venice to Paddington Basin has been the focus of intense profit-driven redevelopment. When the *Paddington Waterside* masterplan was originally approved by the City of Westminster, the plans proposed to replace an 80-acre post-industrial canal-side with a new business and residential district. The facilitation of the development by the City of Westminster and British Waterways followed the national government's policy of deregulation and privatisation, requiring public organisations to release their land assets to private investors and developers. At Paddington Basin, as Julian Dean, a planner for the City of Westminster, described, the 'disposal of [land] assets' by British Waterways led to conversations in the local authority about how to respond (Interview 2013). The subsequent masterplan and resultant developments have been highly facilitated by the Conservative local authority in Westminster, through planning approvals and the establishment of a Business Improvement District (BID), and by strategic transport infrastructure approved by central government.

The redevelopment around Paddington Basin,[1] that has been led by a partnership of developers, has created a frontier of gentrification. It expands through what Neil Smith describes in his research focused on New York City's Lower Eastside, as the 'actions of collective [land] owners' (1996:xviii). The momentum of transformation has been informed by sustained public investment into Paddington as a transportation interchange, through the arrival of Crossrail and

DOI: 10.4324/9781003248187-4

FIGURE 3.1 The large corporate offices and apartment blocks cluster around the canal and the paved canal towpath.

(Photo by author)

the Heathrow Express, along with the privatisation of public agencies National Freight Corporation and British Waterways that once owned the masterplan sites. Central to the masterplan is the canal basin and a network of open spaces that have been built as each of the fourteen development parcels reaches completion. Although the site was once under the responsibility of state agencies, what is being realised is not a public realm like the streets and squares of Central London 'where the streets have been ours since the 1772 Paving Act' (Interview with Westminster planner, Julian Dean, 2012). Instead, this is a site of fragmented property ownerships, management structures and regulations, bringing into question the publicness of the public realm that is being constructed.

In this chapter, I explore the area of the *Paddington Waterside* masterplan, with a focus on the public spaces alongside and interconnected with the canal basin. The unfolding of the large-scale redevelopment, which began in 1997 when the planning team at Westminster began to discuss the area's potential, has relied on the local authority facilitating planning approvals for the developers and collecting rent for the BID. It has also benefited from British Waterways and the National Freight Corporation releasing sites for development and offering favourable leases. The plan can be read as a conventional urban design arrangement of mixed-use buildings, streets, squares, footpaths and an amphitheatre along a rejuvenated canal. But as each phase has been completed, a landscape of well-maintained, high-quality materials has opened up a securitised public realm: the views across the dredged canal, the expanse of granite paving and the reflective building facades are marked by teams of security guards who confront teenagers, homeless people and other individuals who appear out of place. Behind the architect's drawings of Paddington Basin are layers of private controls that result in a lack of diverse or spontaneous activities, concealed disputes between developers and building managers and inconsistencies in claims made by developers, the BID and the local authority.

I investigate how public spaces have been produced in this development west of central London and how descriptions of place-making obscure strategies for private ownership and control. I describe three intersecting narratives of Paddington Basin. First, a physical site of urban development, architecturally designed and constructed. Second, commercial operations of urban development authored to facilitate profit. Between the development of traditional spatial structures of buildings and open spaces and carefully scripted marketing descriptions, I have experienced a process closely directed by private developers who manoeuvre to profit from both the private buildings and the public realm that they create. Third, these narratives contrast with an account of 'place-making', presented by the BID and developer teams: 'It's about people and communities, the surrounding neighbourhood context, and creating a vibrant place to live, work and play', claim the development partnership (www.thisispaddington.com 2017). However, as David Harvey explains, the production of such places through 'architecture and urban design, is precisely about the selling of places' (1996:298). I identify the development of Paddington to be based on the ever-changing ownerships of property, promises of new public spaces and claims by public agencies and private interests to be remaking Paddington Basin as a place. I reveal that behind assertions of place-making, historic structures are demolished, tenants are displaced, and the remaining traces of state-owned assets are handed to private managers. Despite the ambitions of the development teams to generate an identity for Paddington Basin, distinctive places have not been created from the masterplanned transformations. Instead, developers and the BID have incorporated the identity of historic buildings from outside of the masterplan area and refer to fictional characters, such as Paddington Bear, incorporating them into development narratives.

Making space, talking place

> 'Place making is a lot to do with reputation, a lot to do with spirit, and the activities between the buildings and I think that a lot of people get it wrong, and I think it is very hard at Paddington because we have so many different owners.'
>
> —Kate Beaton, Paddington Business Improvement District (Interview 2012)

Narratives of place

There has been an ambition to make a 'place' at Paddington. In 1988 Westminster City Council designated an area of Paddington, then considered on the edge of Central London, as Paddington Special Policy Area (PSPA). The 80-acre PSPA was subsequently named 'Paddington Waterside' by a developer-led consortium, Paddington Regeneration Partnership (PRP). One of PRP's early priorities was the definition of 'place'. An initial document, *Paddington Waterside: Creating a Place*, describes the importance of a 'high quality public realm' that was considered 'vital for improving perceptions and for creating a new sense of identity and place' at Paddington (PRP 2001). The report was published by the PRP to 'achieve consensus' regarding the 'quality, coherence and identity' of the Paddington Waterside development, which included fourteen development sites and twenty-two partners. Richard Jones, a project manager for one of the developers, stated 'the trick with regeneration projects in slightly down at heel areas is to, as quickly as possible, establish a sense of place' (Interview 2013). For Jones this included building the public realm early, or as he described: 'putting the soul into the first phase'. The investment in the landscape of pedestrian streets, bridges, walkways and an amphitheatre along with tree planting, lighting and signage, accompanied the first office buildings that were built over the former goods yards, an area branded 'Paddington Central'.

Through the *Paddington Waterside* redevelopment a new kind of place has been emerging. The Paddington BID, advocated by planners at the City of Westminster and funded by Paddington Waterside Partnership (formerly PRP), began operation in 2005. Interviews with developers, the Canal and River Trust, City of Westminster, the BID and Paddington Waterside Partnership reveal the importance of placemaking to the organisations realising the redevelopment. Harvey identifies that 'those who have invested in the physical qualities of place have to ensure that activities arise which render their investments profitable by ensuring the permanence of place' (1996:296). In addition to the public spaces, the notion of place, as described by project manager Richard Jones, includes the shops under the amphitheatre, the cafes, bars and restaurants along the canal. Creating a place 'requires big investment', Jones stated, which he believes differentiates his development, Paddington Central, from others further along the canal. During the first phases, Jones needed to encourage the retailers of the shop units to remain open despite low sales, promising that once the numbers of residents, workers and commuters reached a critical mass, their businesses would become more profitable (Interview 2013).

The canal basin and the rail networks have been core urban elements in the attempt to define Paddington Basin as a distinctive place. The canal basin marks one of several periods of transportation infrastructure that have been overlaid several times. Originally developing around the meeting of Edgware Road, a Roman Road, and Harrow Road, a Saxon Road, Paddington has been dissected by canals, railways and the infamous Westway motorway that was built in the late 1960s. While the railways have had a resurgence through the Heathrow Express and the Crossrail projects, the canal no longer has any transportation function. It remains active with moored residential boats that move seasonally along the 2000-mile network. The canal basin and large tracts of property that abut the water are owned by the Canal and River Trust, formerly the public agency British Waterways.

The landowners, developers and architects have long recognised the redevelopment value of the land assets. Michael Bond who worked for the Canal and River Trust explained that 'even a small canal frontage . . . could have a 20% impact on the value, perhaps even a little bit more' (Interview 2013), reflecting Richard Jones who stated: 'Estate agents will wet themselves over a view of water' (Interview 2013). This higher land value for waterfront views is particularly marked for residential properties that have increasingly been planned along the canal as development plans have unfolded. Kevin Lynch specifically cites canals in his category of 'paths', one of five urban elements from which he considers the imageability of cities. For Lynch paths are prominent features along which other elements, such as buildings, structures and open spaces, are 'arranged and related'. They are also urban forms, he claims, which have the potential of giving a place definition (1960:47).

The canal, open spaces and land parcels are bound together through their potential for redevelopment. Lynch accepts that several paths can be 'imaged together as a simple structure' and that if there are many paths, such as the towpaths, canals and other open spaces of Paddington, they can be perceived as a 'total network' (1960:59–60). The Canal and River Trust own the canal, its towpaths and many canal side properties. These components were essential in the historic operation of the canal as a water network, which relied on towpaths from which barges would be towed and land for the construction of wharfs and warehouses. As an infrastructure of urban redevelopment, the canal, towpaths and land parcels are also inseparable—the large tracts of land offer a sufficient footprint for profitable development surrounded by water and open space that contribute to increased property value. The open spaces that form the new public realm add 'real value' to the development, as Michael Bond, of the Canal and River Trust, explained. He elaborated how they had 'worked' their portfolio of properties as a major source of income: 'we have our growth portfolio . . . which is all about effectively generating capital receipts and capital value, normally through the planning process' (Interview 2013). At Paddington Basin this involved selling a 999-year lease to a development partner and sharing the profits generated through its transformation. The redeveloped canal edge, the towpaths and the development parcels are now almost indistinguishable from each other suggesting that, rather than the canal on its own, they collectively act as an armature or network across the development area.

Paddington Basin is becoming a particular type of place. In the *New Urban Frontier* (1996), Smith describes urban transformations through gentrification and displacement. The frontier, he claims, 'adapts to place as it makes place' (1996:16). Although the term 'frontier', to which Smith refers, is more closely associated with North American colonisation, there are correlations with the redevelopments at Paddington Basin, Elephant and Castle and others across London. In 1987 Julian Dean and his colleagues at the City of Westminster's planning office recognised Paddington's potential as an outlying area of central London ready for redevelopment. Paddington was considered just outside of London's 'central activity zone'. Dean explained that 'all the headquarters of government, state, crown, church,

FIGURE 3.2 Smokers stand outside the office buildings while there are limited places to sit along the canal-side of Paddington Basin.

(Photo by author)

shops, thirty-eight theatres, cinemas, all human life' were inside the central activity zone; in contrast, outside the central activity zone was Paddington, a district that had suffered for decades from areas of poor housing and a decline in railway and canal infrastructures across an area that still revealed the traces of bomb-damage from the Second World War (Interview 2012). As Smith describes of the urban frontier, Paddington Basin was considered by the developers as a 'hostile' landscape waiting to be 'regenerated' (1996:12), a place that Richard Jones confirms was 'pretty rough' before the development began. Across sites like Paddington Goods Yard 'if you weren't tripping over rusty old car parts you were tripping over prostitutes and needles' (Interview with developer, Richard Jones, 2013).

The first stage of creating a new type of place is the closing down of the remaining businesses and the removal of buildings. Harvey attests that: 'Old places . . . have to be devalued, destroyed, and redeveloped while new places are created' (1996:296). Jack Whitehead, the author of the website *Local Local History*, describes his experience walking along the Paddington Basin in 1999: it was a place 'totally demolished' where buildings had been bulldozed to clear the site which 'would become the anonymous foundations for new buildings' (www.locallocalhistory.co.uk 2012). Although Westminster planner Julian Dean claimed that before the development there were no major buildings along the canal (Interview 2012), Whitehead evidences a 200-year loading bay at the head of the canal that had been listed for its historical interest. The loading bay, like the bridges designed by Brunel, were dismantled and put into storage to facilitate a site clear of obstructions from which the masterplan could be actioned. The resultant tabula rasa was efficient for establishing the large floorplates that corporate tenants required across evenly subdivided, uninterrupted development parcels. As Jane Jacobs recalls from redevelopments in New York, the 'landmarks' across the masterplan site were 'crumbled or sundered from their contexts' and the new place that emerged could only be described as a 'noplace' (1961:339).

The *Paddington Waterside* area was cleared of its historical buildings by the time the BID was established in 2005. Therefore, new landmarks and points of reference were sought beyond the masterplan site. Kate Beaton, leader of both the BID and Paddington Partnership (formerly Paddington Waterside Partnership and before that Paddington Regeneration Partnership) described that there was a 'beautiful heritage about Paddington' that the BID was keen to promote (Interview 2012). Despite pessimistic descriptions of the deteriorated industrial sites, poor housing and peripheral location claimed by Julian Dean (Interview 2012), beyond the development parcels there were architecturally significant structures like the railway terminal, the Great Western Hotel and many magnificent terraces of Georgian houses. Beaton explained: 'We wanted to take some of the flavours of the station, because the station is so—there is a romance about Paddington Station' (Interview 2012). The BID area therefore encompasses the historic streets to the west and south of the masterplan area and its marketing material strongly references the historic station. Recognising the potential of the deteriorated margins of the city, Smith reminds us that these places are not only unsightly and dangerous, but they are also idyllic—'romantic but also ruthless' (1996:12). Between the BID and the development partnership their websites reference an array of locations inside and outside their bounds, including landmarks such as Paddington Station and Hyde Park and adjoining neighbourhoods such as Maida Vale. Unable to benefit from the historical assets removed from the development area, the marketing teams need to look further afield to author a new account of place at Paddington Basin.

Replacing the close-grained intricacy of the industrial wharfs, developers, the local authority and the Canal and River Trust have realised a high-quality, intensely managed, consistently paved public realm that wraps around modern glazed and brick facades of private residences and businesses. Despite marketing narratives of Paddington as a unique place, architecture critic Rowan Moore writes in the *Architect's Journal* about an increasingly popular model of development that he feels has been adopted at Paddington Basin (www.architectsjournal.co.uk 2009). The 'essential aspects' of this model, Moore writes, include 'efficiently planned flexible blocks, which give corporate tenants the floorplates they want' alongside the 'creation of high-quality spaces between buildings'. Moore denies that the external spaces are public spaces because of the private management, but he recognises the value of the 'high specifications of paving materials, coordinated street furniture, and outdoor art.' Around the buildings of each development parcel are squares, terraces and steps of granite, stainless steel balustrades and hardwood furniture alongside occasional groups of semi-mature street trees. This is a palette of materials that can be found across many sites in London that Moore claims follow the 'Broadgate model'. He describes that 'places like Canary Wharf, More London, Paddington Basin and the Regent's Place development on Euston Road, all follow its pattern' (www.architectsjournal.co.uk 2009).

The undifferentiated buildings and open spaces that have been produced through the redevelopment of Paddington Basin, a site that now resembles many other developments in London, contrasts with developers' interests in defining a place of uniqueness. Paddington has come to be seen to represent 'pretty much the mainstream', according to architecture critic Kieran Long, 'in terms of how we do go about making new bits of London today' (2010). In contrast to the smooth paving around the development parcels, along the edge of the canal traditional cropped granite paving setts provide what Michael Bond from the Canal and River Trust describes as the 'cobble effect' (Interview 2013). This historic technique of paving, where stone masonry provides roughly cropped stone, is a requirement of the Canal and River Trust to give a heritage feel and continuity along its infrastructure of waterways. The towpaths along the canal that form the main public rights of way through Paddington Basin continue to be owned by the former public agency. The rough paving contrasts with the machine cut granite paving that is laid in the development parcels. Although it may invoke a sense of a more historical place, Bond feels that the traditional paving technique used on the towpath is one of the 'biggest mistakes' of their involvement in the development, as 'it's difficult to walk on in heeled shoes' (Interview 2013). Despite the heritage materials specified along the canal, the management, security and maintenance of the public towpath at Merchant Square are not conducted by the Canal and River Trust, but rather contracted out to the private operator, Broadgate Estates, after whose 'model' Moore claims the Paddington Basin development has followed (www.architectsjournal.co.uk 2009).

The *Paddington Waterside* masterplan also seems to follow Lynch's commonly accepted planning model for imageability (1960). The masterplan encompasses a scale of London that Lynch would describe as a 'district' (1960:66). Some of Lynch's other urban elements for creating a 'city image' are also identifiable within the development: 'paths' follow the canals and pass between buildings; 'nodes' and 'landmarks' are suggested through the footbridges that cross the canal, however, their small scale seems overshadowed by the large buildings; while 'edges' are also difficult to discern, especially at the extent of the development area and in between the development parcels. The significance of image making is emphasised by developer Richard Jones, who considers place-making as an 'aesthetic investment' (Interview 2013). Through offering high-quality spaces, punctuated by bridges and artwork as landmark features, the developments follow the priority of the policymakers at Westminster who have attempted to 'change the image of the area' and to shift the perception of Paddington from a district on the outside edge, to a 'central place' in London (Raco and Henderson 2009:311). Lynch warns, however, where paths lack identity or are easily confused with one another the 'entire city image' can be 'in difficulty' (1960:49). The similarity of the high-quality open spaces in Paddington Basin with other contemporary developments across London, along with the unbroken continuity of open spaces along the canal, highlight the difficulty in forming a clear identity of a new place.

Property and space

Despite ambitions to make a place, the reality of Paddington Basin has been the making of space. Mike Raco and Steven Henderson suggest, the development has transformed the area from 'a problem place to an opportunity space' (2009:305). By 2013, over two million square feet of commercial space had already been 'delivered' (www.paddingtonwaterside.co.uk 2013) across a development that initially included over eight million square feet of 'built environment' (PRP 2001). Of the residential units for sale, the marketing brochure for 3 Merchant Square boasted apartment sizes that ranged from under 500 square feet for a one-bed apartment to up to 1377 square feet for a three-bed (www.merchantsquare.co.uk 2013). The developer presented options of combining these spaces to create apartments with up to seven bedrooms across an area of 3635 square feet. The apartments for sale at 3 Merchant Square claimed to offer a 'flexible living space' while apartments in 1 Merchant Square were described as 'perfectly proportioned' with a 'sense of space' (www.merchantsquare.co.uk 2013). The interchangeable use of the term space and place is touched on by Ali Madanipour, who explains that 'the meaning of the two concepts [space and place] often merge' (1996:23). Attempting to differentiate between the terms, Tim Cresswell describes space as a 'fact of life' rather than place, which is vested with meaning: 'Space, then, has been seen in distinction to place as a realm without meaning' (2004:10). The marketing of space offers residents an opportunity to give meaning to the apartments that they purchase, allowing them to redefine their properties as places. For investors, viewing the purchase of an apartment as an investment, the term space frames a finished product, a neutral commodity that can be traded. Like the development parcels across Paddington Basin cleared of their historic buildings and uses, enclosed in timber hoarding and traded between developers, the apartments offer an assurance of the value of space.

As new buildings are sold in terms of their spatial dimensions and qualities, the sense of place that developers seek is focused on the features that compose the external realm. Open spaces, managed by the developers and the BID, which are surrounded by bars and restaurants, along with visual associations with historic streets, buildings and landmarks, are employed to make Paddington Basin a meaningful place. However, what is being made at Paddington could also be described as a 'non-place' (Augé 1995:94). Commonly associated with airports and shopping malls, Marc Augé describes 'non-places' as 'spaces formed in relation to certain ends', like transit, transport, commerce and leisure, and defined by the 'relations that people have with these spaces' (1995:94). Despite the marketing narratives of place-making, Paddington Basin can be seen in the context of the Broadgate model: a commercial development that is insufficiently differentiated from other areas of London with objectives of producing a financial profit. The imperatives for economic return are reflected in the operations of both development partners, the BID and the former public agencies. In the context of the high-quality but poorly activated public realm, the relations that many people have to these spaces is bound up with their workplace or the economic value of the land they own and the apartments in which they reside (Interview with residents, Edward Shoemaker and Sharon Peters, 2012).

Movement and obstructions

> 'You go past [Paddington Basin] without even knowing that it was there'.
> —Peter Moorland, architect and masterplanner (Interview 2013)

Movement facilitated

Most people easily access Paddington's new waterfront spaces. Access into this central London regeneration project, as defined by the development masterplan and the BID, appears consistent and evenly applied. There are over twelve entrances into the Paddington Waterside site, from adjacent streets, towpaths and public rights of way. The development parcels have been required to provide accessible open space, through Section 106 agreements (Town and Country Planning Act 1990) and guidelines from the Canal and River Trust. Paddington Basin is the result of a masterplan that claims to open up the formerly inaccessible and neglected canal side areas to create a new, dense, waterfront, mixed-use neighbourhood surrounded by permeable open spaces (Interviews 2012, 2013). Whitehead recalls that the masterplan proposed by Terry Farrell and Partners in 1999 claimed that the canal-side area designated for redevelopment 'lacked public activity, permeability and a sense of place' (www.locallocalhistory.co.uk 2012). For two centuries Paddington Basin had been an active waterfront, with limited public access and a clustering of industrial buildings along the canal. In contrast, the development prioritised accessibility and permeability with the canal towpath providing an armature of connectivity. The new developments open onto, are organised by and are served by canal-side paths that weave from one side to the other across a series of pedestrian bridges. The towpaths connect to the streets and spaces beyond the canal basin suggesting a permeable environment throughout. Subtle changes in materials and linear drainage gullies indicate thresholds of ownership and management, or boundaries of the phased construction. However, there are no gates, walls, fences and few changes in vertical elevation to separate the towpath from the development sites.

Despite the accessibility of the area, Paddington Basin has been uncomfortably quiet, particularly in the evenings and during the weekends. Until the Canal and River Trust and the developers initiated a programme of events and activities, such as boat hire, a water taxi, and floating cafes, most people tended to keep moving through the area with minimum engagement with each other or the spaces across which they move. Even with these efforts, and beyond warmer days during the summer, the exposed canal-side remains windy, shady, and unwelcoming. Peter Moorland, an architect of one of the development parcels, expressed concern for the masterplan as a 'completely introspective piece of work' (Interview 2013).

FIGURE 3.3 The canal-side open spaces have increasingly been appropriated for use by the commercial tenants inside the buildings.

(Photo by author)

Although Moorland expressed that this should improve as the development is completed, the design makes it difficult to identify the front of the development from the back, or where people should arrive or leave. The inward-looking design of the development contributes to a lack of permeability along edges that are blocked by buildings or dissected by the railway or the canal. It is difficult to exit Paddington Central to the north as the change of levels and the concealed access makes 'you feel as if you shouldn't be there' described one of the developers (Interview with developer, Richard Jones, 2013). Jones explained: 'It is not physically impossible [to go through our estate] if you know where you are going, and the security guards won't stop you' however he recognised that the condition of the spaces makes 'people feel that they are trespassing'.

Throughout the development, routines of cleaning, maintaining and securing the open spaces remind visitors that this is a very different form of public realm to that which exists beyond the masterplan area. Despite some developments being completed over ten years ago, there are few marks of aging, use or past inhabitation. Paving is swept, windows are cleaned, trees are pruned, handrails are polished and the canal is dredged of algae. Throughout the day and night, the building developments and the open spaces along the canal are maintained to an extremely high standard. The cleaning routine is only interrupted by the need to replace broken or damaged materials or the events programme that hosts family and business-focused activities one day, to leave no traces the next. Cresswell states that when creating a space, it is necessary to be explicit about what lies outside (2004:102). The main difference between Paddington Basin and areas beyond the development masterplan is that Paddington Basin appears to be permanently new. Through intense management, policing and cleaning the public realm reveals no physical traces of what has happened on the days, weeks, seasons and years past. Smith explains that 'new urban pioneers seek to scrub the city clean of its working-class geography and history' (1996:25). With old buildings demolished, former land-uses extinguished and a management regime that holds the new forms and activities in place, Paddington Basin can be understood as a different type of place and public realm than exists outside the development and BID boundaries.

Amongst the people who visit and pass through Paddington Basin, teams of private security guards are omnipresent. Young and middle-aged men in black and white uniforms, some additionally wearing bright yellow jackets, circle the development parcels enforcing the various rules written by the developers. Extending beyond the five spatial urban elements that he describes for achieving image and identity in the city, Lynch describes in *A Theory of Good City Form* (1981) the importance of 'performance dimensions' that are not so easy to determine. Lynch includes 'control' as a measure of performance, asking how 'variations in control' can affect the spaces of the city and explaining that 'we are accustomed to one particular form of control' which is the 'legally defined ownership of a sharply defined area' (1981:205). However, at Paddington Basin, as the canal basin and adjacent streets provide almost unbroken across the development parcels, ownership and property boundaries are not easily differentiated. Also, control is not only exerted through the security personnel, but instead through the collective presence of cleaning, maintaining, security and events. Together, the four operations facilitate the continued flow of people through the area. The recognisable architectural forms that have been built across this 80-acre site for almost three decades are held in time by maintenance, while the flows of people passing through are highly managed by a schedule of development operations.

Obstructing engagement

Paddington Basin's developers and BID aim to control public spaces, to hold the area *in place* and maintain its image. This results in an open and accessible public space coming into conflict with what Don Mitchell describes as 'urban aesthetics' (1997:306). Mitchell finds an increasing occurrence of homeless people excluded from public spaces through what Sharon Zukin describes as the manipulation of 'symbolic languages of exclusion' and 'uses of aesthetic power' (1995:7). The developer operations of cleaning, maintenance, security and events also closely reflect the priorities of the BID: 'environment, safety and security, and the marketing' (Interview with resident, Sharon Peters, 2012). The main three development areas, West End Quay, Merchant Square and Paddington Central, employ their own teams of security, maintenance and cleaning staff. At West End Quay these operations overlap with the actions of the BID. While it can be understood that a commitment to place-making facilitates planning approval and that attention to space is fundamental for investor interest, once the latter conditions are satisfied a requirement of control across the open spaces is important to maintain the high-quality investment and the resultant image of the area. The cleaning, maintenance, security and events hold the image of the development in place, attempting to recreate the architectural renders that initially facilitated planning approval as a means to attract tenants and investors. These are images that represent 'a place of comfort, of relaxation perhaps, of leisurely consumption, unsullied by images of work, poverty, or social strife' (Mitchell 1997:323), aiming to ensure the continued and long-term income that the Canal and River Trust and their investors expect.

The operations of cleaning, maintenance, security and events are also profitable businesses in themselves. Following their success at More London, Paternoster Square and Liverpool One, management company Broadgate Estates was subcontracted to manage the development at Merchant Square. Claiming to help 'manage and build the value of [their] clients' assets', Broadgate Estates manages completed developments, like Merchant Square, for profit. They claim that their public realm estate management provides the 'highest standards of cleaning and maintenance', opportunities for 'commercialisation' and services that can 'enhance public life' (www.broadgateestates.co.uk, 2014). These operations transform the priorities and the perception of the public realm, creating at Paddington Basin what Kieran Long describes as 'a weird place, ostentatiously guarded by private security' (Long 2010).

The activities of private management at Paddington Basin make it difficult to recognise the open spaces as public. Westminster City Council intended full public access to the canal and basin and have set this out in their Replacement Unitary Development Plan (RUDP) and their Local Development Framework (LDF). *Paddington Waterside: Creating a Place* promotes towpaths as unbroken public routes through the area, facilitating access, permeability and legibility between the developments—and being 'pivotal' in achieving the 'sense of place' (PRP 2001). However, the public nature of the spaces is difficult to decipher. Most of the open spaces at Paddington are described as public spaces by the local planning authority and by those designing and managing the external realm. The public spaces can be understood as the common spatial types of squares, streets, and towpaths, but there are contradictions in how the spaces are perceived, experienced and proposed as public spaces. The private activities of cleaning, maintenance, security and events that are imposed on Paddington Basin make it difficult to engage in shared, collective and public actions. Behind the simplicity of spatial forms, many questions remain about the claims of creating public spaces, mixed use developments and community engagement by those responsible for the development.

Instead, Paddington Basin produces a different public. Other than walking through Paddington Basin, the activities curated across the new waterside are mainly commercial ventures of cafes and bars—and what is described as an Artisan Market. The open spaces around each building development, that were designed as part of planning agreements, have been increasingly used as expanded seating areas for the restaurant and cafe chains inside the buildings as well as sites for food trucks and vendors. Even the canal has been occupied by a series of canal boats converted into bars and restaurants. Recognising the lack of activities around the development areas the developers and the Canal and River Trust have focused on business-related activities that produce a different public life than that experienced on the other side of the development on Edgeware Road. But rather than denying the existence of a 'public' at Paddington, we can instead recognise an 'exclusive public' that is 'produced and managed by narrow interests' (Madanipour 2010:10) and whose spaces are animated by individuals involved in the commercial events, security, maintenance and cleaning.

The open spaces of Paddington Basin can be easily entered, but movement across them is not guaranteed. The openness facilitates, as Lynch describes, 'a right of presence' (1981:206). Lynch sets out five 'rights' that comprise control in cities that extend beyond legal terms of ownership. At Paddington Basin this first right of presence is ensured through public access afforded along the canal towpaths. Michael Bond of the Canal and River Trust explained, 'we have to provide public access to towpaths at all times'. The adjacent spaces that have been committed to the public as part of Section 106 planning agreements also allow public access, reflecting Madanipour's assertion that 'without being accessible a place cannot become public' (2010:7). However, the conditions of access to Paddington Basin can easily be questioned when 'use and action', the second of Lynch's rights, are restricted (1981:206). Polished wall mounted stainless steel signs remind those passing through that they are on private property and that there are certain prohibited activities. The large number of different signs at Paddington Basin, which proclaim that the 'public spaces' proposed at planning are actually private property, outlines the main activities that are unacceptable. These include a diversity of notices proclaiming the prohibition of smoking, skateboarding, rollerblading, scooting, cycling, feeding pigeons, unauthorised parking, double berth mooring, trespassing and even public access. The signposted regulations are enforced for the landowners by the private security guards who further restrict taking photographs of the buildings and who question the presence of groups and individuals who may appear out of place.

Private security guards are tasked with enforcing what is acceptable within the development areas. Cresswell affirms that places are 'created by some people with more power than others to define what is and what is not appropriate' (2004:27). Paddington Central and Merchant Square have separate security teams from those of the BID, which overlaps with security at West End Quay. The BID pays the Metropolitan Police for additional officers and contracts out CCTV surveillance. Richard Rogers, whose architecture firm designed one of the buildings at Paddington Basin explains the potential problem of privately owned and securitised public spaces:

> Private security forces patrol their precincts and decide who is not welcome. People who simply want to sit are treated with suspicion, and groups of lads are frequently banned 'as a precautionary measure'. We are witnessing a new generation of enclosures, which may have an effect as long-lasting as those of the eighteenth century. Now, as then, people's right of access to public space is being taken away.
>
> *(Rogers and Fisher 1992:111)*

The security teams in the development area have been criticised for what Julian Dean from the City of Westminster described as being 'bolshy'. Dean explained that this is 'because we live in strange times' and accepts 'that is pretty much what we have got' (Interview 2012). Michael Bond from the Canal and River Trust elaborated: 'people these days are actually paranoid, in particular about terrorism' (Interview 2013). At Paddington Central, where large and occasionally controversial corporations reside as tenants, there is also a concern for privacy from protestors. Bond described how animal rights protestors 'turned up' to demonstrate against the pharmaceutical company, AstraZeneca. The small group of activists were aware of the different divisions of land ownership and stood on the towpath, a public right-of-way owned by the Canal and River Trust. Bond recalls that 'they knew their stuff', in relation to the nuanced landownerships, which meant that 'the security people at Sheldon Square were doing their nut' (Interview 2013). Richard Jones, who oversaw part of the development of Paddington Central, emphasises that some of the corporate tenants are requesting these levels of security and privacy, a view supported by Bond who offered the 999-year lease for the Merchant Square site to European Land: 'I think that is just the way of the world and you can't blame the property developers for that' (Interview 2013).

Through the masterplanning process, the developers and the council have managed and designed-out people and uses that they feel are incompatible. Before the North Westminster Community College was closed and demolished to make way for the development, security guards at Merchant Square frequently confronted young people to deter them from congregating, drinking and smoking at Paddington Basin. The BID routinely 'sends the police in' to disrupt 'young lads' hanging around in the canal-side spaces, behind Tesco and Ladbrokes (Interview with development and BID manager, Kate Beaton, 2012), while Westminster City Council evicted a homeless man and dismantled his shelter on Stone Wharf. Lynch advocates that 'the management of all but very small areas must be tempered by the participation of potential users elsewhere' (1981:208). The participation of young people has been diluted by the closure of the college while the presence of homeless people has declined since many hostels and shelters have been closed following the gentrification of surrounding streets. Where the 'redevelopment frontier advances', Mitchell and Staeheli explain, it is increasingly difficult to find buildings and spaces for homeless residents (2006:167). As neighbourhoods are developed and gentrified, 'public spaces remain the places where homeless people can be' despite all the changes that redevelopment brings. However, as the school and shelters were closed across Paddington, teenagers and homeless people appear increasingly out of place in the gentrified neighbourhoods. Rogers and Fisher claim that 'the steady erosion of diversity of activity has been an unpleasant side-effect of *laissez-faire*, market-led development' (Rogers and Fisher 1992:xxv). This resonates with Julian Dean who explained that 'it is not an act of policy', at Westminster, but rather 'how the market works', concluding that there are other places in which homeless people can 'live more easily' (Interview 2012).

Repackaging Paddington

> 'We are all about creating destinations. Attracting people to our network. One of our KPI's is visitors, the number of visitors. And by visitors we don't just mean people in boats. We mean walkers, joggers, cyclists, dog walkers'.
>
> —Michael Bond, Canal and River Trust (Interview 2013)

Commodifying

The Paddington Waterside development, and the wider area of Paddington under authority of the BID, is regularly repackaged. Harvey describes that during 'speculative place construction' through investment in urban development 'the selling of places and the highlighting of their particular qualities . . . becomes even more frenetic' (1996:298). At Paddington Basin, continual rebranding and marketing facilitates residential and commercial development, which are measured in terms of financial return. The developers boast about the metrics of space being produced: they describe the increased number of floors, the area of large floorplates in their new office buildings and the quantity of bedrooms and bathrooms along with the square footage of apartments. The marketing narratives of the BID present what BID manager Kate Beaton described as the 'softer side' of development, including the less tangible and elusive notion of making a 'place' (Interview 2012). These conditions supported by the BID are further encouraged by the Canal and River Trust which focuses on attracting people to their canal network, directly reflecting Harvey's assertion that:

> It is also about attracting consumers (particularly the affluent) through the creation of amenities such as a cultural centre, a pleasing urban or regional landscape, and the like. Investment in consumption spectacles, the selling of images of places, competition over the definition of cultural and symbolic capital . . . all become conflated in inter-place competition.
>
> *(1996:298)*

Throughout the development area there has been a continual division and subdivision of ownership and management responsibilities, resulting in the area experiencing a fragmented identity. Large development parcels have been divided, new companies have bought into the development and management of the developments fragmented. As a result, a constant multiplying of names, brands and sub-brands has occurred beneath the three main organisations of Paddington Partnership, the Canal and River Trust and the Paddington BID. Kate Beaton described the problem with place-making at Paddington Basin as the development parcels have been traded between developers and the identity became complicated: 'I think it is very hard at Paddington because we have so many different owners'. Beaton elaborated:

> One of the developers at the time, then called Chelsfield, then Paddington Basin Development, then Paddington Basin Development Corporation and then European Land, owned the same bit of land but went through four iterations of name. Then Paddington Central, that was Regalian, which sold to Development Securities, no, not Regalian. Yes. Regalian, who had a joint venture with NFC, National Freight Company, at the time, and they sold to Development Securities who bought it speculatively.
>
> *(Interview 2012)*

The first identity of the development across Paddington Basin has been that of the development partnership, Paddington Regeneration Partnerships (PRP), later renamed Paddington Waterside Partnership (PWP), and then Paddington Partnership. Initiated in 1998 by Chelsfield, one of the major developers, the aim of the PRP, as Chief Executive Kate Beaton describes, was to provide 'a coordinated development platform' (Interview 2012). The eight founding partners included private businesses and several public organisations, the latter of which, through government policy over the previous decades, had been gradually privatising or commercialising their activities. Despite the competing interests, who have since sub-divided, planned, commodified, sold and resold their long leaseholds to one another, the sharing of development information has facilitated a coordination of approaches and standards. The partnership provided services beyond its own boundaries and brand, *Paddington Waterside*. It operated a recruitment service, *Paddington First*, that encouraged employment in the larger area; PWP also assisted companies to deliver corporate social responsibility programmes through a separate organisation, *Time For Paddington*; and it led and operated the *Paddington BID* (Business Improvement District). Together with the BID, it created the identity of *InPaddington* which identified the value in tangible and ephemeral elements of the area, such as of Paddington Bear, Isambard Kingdom Brunel, Hyde Park and the Royal Family, to promote itself and the Paddington area. This strategy by the partnership (with the BID) advanced what Setha Low explains as necessary for the 'takeover of the public space of entire communities and neighbourhoods' (2006:83). Through this suite of companies and structures, the partnership went beyond the appropriation of public spaces—it subsumed and commodified cultural and historical figures, fictional characters, spaces beyond Paddington and even the smaller private interests on the edge of the development area.

The emerging identities have been the Paddington BID, branded as Paddington Now, and the Paddington Partnership. The Paddington BID was initiated by the partnership in 2004 and was set out to extend beyond the development boundary. Kate Beaton, the BID CEO, described that the role of the BID was to capture the benefits of the regeneration and Section 106 contributions through 'economies of scale' which achieve 'more bang for our [the partnership's] buck'. She explained that the BID aimed to benefit from the 'softer . . . social, community and economic' activities that can be understood to contribute to making a new place (Interview 2012). Sharon Peters, a local resident on the board of the BID, explains how the BID provided an additional level of service above what was provided by the council (Interview 2012). The concept of BIDs was imported from the US, where they have been criticised for privatising urban areas (Zukin 1995) and have been adapted through the Local Government Act 2003 that describes them as a 'partnership between a local authority and local businesses' (www.parliament.uk 2021). Sharon Peters, however, underlined that the local authority is 'not in control of the BID'. Instead the BID is considered to have a closer relationship with the development partnership (Interview 2012). Kate Beaton who manages both the BID and the Paddington Partnership, stated that the two organisations are operationally and financially entwined: 'it would be very hard to separate the two'. The key requirement of any BID is that it must be voted in by the majority of businesses in the area; once voted in, all businesses within the boundary must pay a levy towards its operation. In Paddington, in 2013, the total raised by the levy was approximately £600K a year. Although the development partnership provided financial support in the establishment of the BID, Kate Beaton recognised that 'there are enormous benefits to the Partnership of the BID, because it brings additional resources to market the location and to improve the location' (Interview 2012). The success of the Paddington BID has since led the Paddington Partnership to establish the Marble Arch London BID (2016), an area that encompasses Edgeware Road, and Hyde Park Paddington Neighbourhood (2014), each extending the influence of the Paddington Partnership across commercial and residential areas of West London.

Through the BIDs, large developers have succeeded in extending their control of the public realm to include neighbourhoods and landmarks beyond the *Paddington Waterside* masterplan area. Neither the purchase and development of the land from British Waterways and the National Freight Corporation nor the management of the canal towpaths provided the Paddington Partnership with a way of establishing a strong image of the development area. Therefore, the BIDs were established to slightly overlap with the development masterplan, and to extend the development 'frontier' into the adjacent neighbourhoods. Harvey explains that 'gentrified neighborhoods' as those emerging beyond the development area of Paddington Basin 'arise on the frontiers of capitalist development or out of the ashes of deindustrialized communities' (1996:296). The developers required additional means to control the neighbourhoods beyond the masterplan area (Low 2006), so established and part funded the BIDs. Without the Paddington BID,

Kate Beaton explains that the image of the development could 'fall off the edge of Paddington into a different Paddington' (Interview 2012). The BID is responsible for cleaning and waste collection, it hires additional police and community support officers and it coordinates with local businesses. The BID is able to direct the public spaces and the private businesses, extending what Harvey terms the 'commodification of everything', including 'cultural forms' and 'histories' (Harvey 2005:160). The coordinating brand of *InPaddington* was created by the development partnership and the BID as a marketing umbrella that co-opted real spaces and historic narratives into the control of the two private organisations.

To control the Paddington BID area, which includes thirty-four streets across Paddington, is expensive and the developers and their corporate tenants have not been willing to pay the full cost. When the Paddington BID was established, its boundary fitted with neither the masterplan area nor what Westminster Council defined as Paddington. Instead, the BID area was extended beyond the masterplan boundary, and only occasionally included any of the development parcels. The large developments in which the international businesses reside, namely Merchant Square and Paddington Central, were notably omitted. Because the BID levy is based on the rateable value of the businesses, the larger and more valuable corporations that were attracted to the large floor plates of the new buildings would contribute larger amounts than the smaller businesses on streets outside the development, like Praed Street. Kate Beaton explained that there are costly management charges already for Merchant Square and Paddington Central: 'whether they [the developers and tenants] pay on top of that to have something that they are already paying for' she said 'is too big a risk to take'. She described that it was initially intended that the large businesses along the canal would benefit the local area. The BID thought that corporate employees would enter the local community cafes, shops and restaurants at lunchtime and at the end of the day (Interview 2012). This has not transpired as the large buildings that host these corporations provide subsidised canteens to accommodate their employees and there are increasing numbers of restaurants along the canal. The need to exit the enclave of the development area has been further diminished now that the train and tube station open up to the canal-side allowing commuters to get directly to their offices bypassing all local neighbourhoods and businesses.

The complicated management responsibilities at Paddington Basin and attempts to keep costs down while maximising their return has brought developer organisations into conflict. The most significant has been a lengthy court battle over contradictory management responsibilities and service charges. Kate Beaton described the complexity of the court-case, but finally joked, 'I still don't know who won' (Interview 2012). The case was between two developers who had at different times previously owned West End Quay. In 1995 when the Paddington Development Corporation (later European Land) signed a 999-year lease with British Waterways, the lease included several clauses regarding management procedures and charges. However, by the time Paddington Development Corporation sold phase 3 to another developer, these procedures were not in place, and new owners established

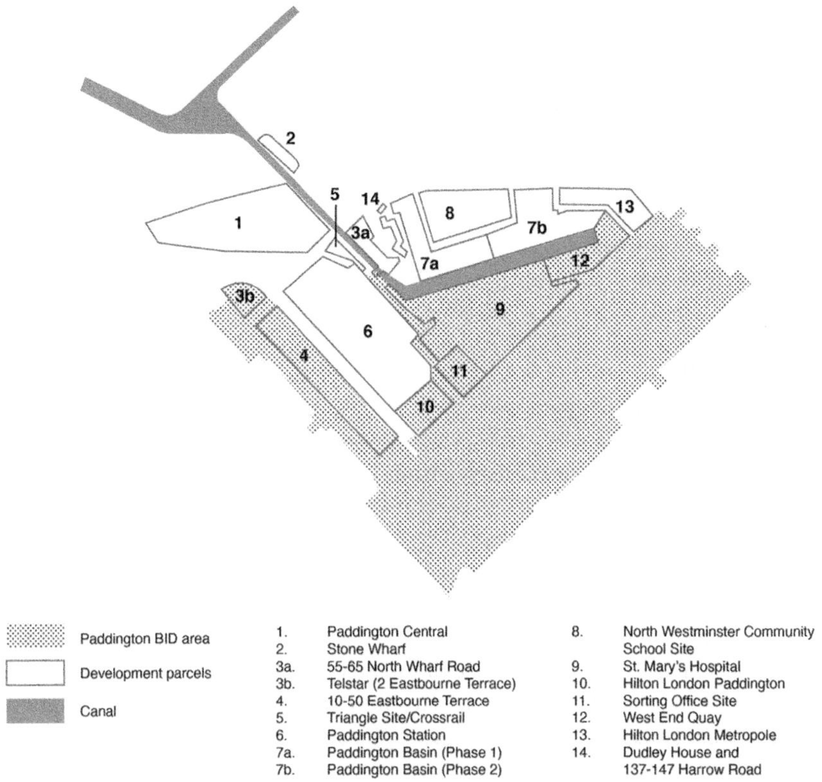

Map based on information from Paddington Waterside Partnership (2009) and Paddington Now BID (2013)

FIGURE 3.4 The Business Improvement District (BID) that is supported by the development partnership extends into the surrounding neighbourhoods.

(Map by author)

their own management responsibilities and charges. Over the following years further management and ownership companies were formed by these two developer groups, all of which had an interest in the land. Ultimately the residents of West End Quay were paying several different service charges and a legal case ensued.

With a proliferation of marketing identities and objectives from the individual developers, the partnership and the BID, incoming residents have not always received what had been promised. Sharon Peters, who bought an apartment off-plan, described how 'the image that was being created' by the developer's marketing teams was 'what I would call a lively riverside, waterside, with boats in the basin [and] with shops and cafes around the side' (Interview 2012). What has emerged in the intervening years 'has been quite significantly different, in some ways, to what was proposed' (Interview 2012). In several interviews with individuals involved in the redevelopment, it was suggested, as Michael Bond of the Canal and River Trust claimed, the difficulty of place-making was that there was not yet a 'critical mass' of people and activities (Interview 2013). However, I would argue that the BID's priorities of environment, safety, security and marketing contributes to overly controlled and marketed open spaces. The 'lively' spaces promised to incoming residents are subdued and sanitised by the aspirations of securitisation: 'The result is that places that seek to differentiate themselves as marketable entities end up creating a kind of serial replication of homogeneity' (Harvey 1996:298).

Conclusions

> Another area of London's public domain that should be developed is its network of canals. . . . There are several sites with considerable development potential, such as Paddington Basin.
>
> —(Rogers and Fisher 1992:121)

In this chapter I have highlighted competing definitions of place, instrumentalised through processes of privatisation focused on the public spaces of Paddington Basin. First, I have recognised that the masterplanning activities at Paddington Basin have been the primary stage of a relentless process of privatisation facilitated by public interventions: the development process was initiated by the sale of formerly public land assets by British Waterways and the National Freight Corporation; it was intensified by the promise of central government to provide strategic Heathrow Express and Crossrail infrastructures, that when built, provided direct access into the development areas; it was made possible by planning approval being given to all development proposals by Westminster Council; and it was expanded through the encouragement by local government for the developers to establish the BIDs. These actions of public agencies were followed by intense developer speculation, an expansion of the influence of the developers through the establishment of the BID and an influx of smaller-scale investors buying property outside the development area. This upgrading of the wider Paddington area has had significant impacts. Such patterns of upgrading are recognised by Smith, who describes that, where 'urban pioneers' such as individual property owners and businesses 'go bravely forth, banks, real estate developers, small-scale and large-scale lenders, retail corporations, the state, have generally gone before' (1996:xviii).

Second, although it could be expected that some buildings within the development area would be demolished as the masterplan is realised, a displacement of undesirable activities has unfolded. The police were employed to remove teenagers from the development area and the community school was closed and relocated. Beyond the masterplan area, prostitution and kerb-crawling along Praed Street are less evident than they were before while homeless shelters were closed down as the value of properties in which they once resided rapidly increased. Sharon Peters, resident and Paddington BID board member, explained that the developer-led partnership facilitated the BID so that the area around Paddington and along Praed Street could be 'regenerated to complement the new development and improve the whole area' (Interview 2012). In this way, the influence of the BID in improving the environment, security, safety and marketing extends inside and outside of the masterplan area resulting in what Julian Dean of the City of Westminster recognised as, 'social cleansing' (Interview 2012):

> Because of the dynamics of central London policy, the degree of what some social commentators call social cleansing does go on. It is called how the market operates, so you might as well call it for what it is. It's not an act of anyone's policy.

The third conclusion is that narratives of place-making obscure a focus on ownership and property. Through the period of redevelopment, a constant rebranding has occurred, evident through the construction and reconstruction of websites for the developments, masterplan and BID. As the masterplan nears completion Julian Dean emphasised that any 'degree of finish' will only be achieved 'on the basis of land ownership' (Interview 2012). He observed development parcels being repeatedly subdivided, sold and bought, with each subsequent owner attempting to increase the value of the land, apartment or property. Financial interactions continue to complete a development area of valuable, sellable spaces. Julian Dean believed that what was left was a matter of 'stitching together between the schemes as they present themselves' (Interview 2012)—a need for 'place-making' by Paddington Partnership and the BIDs. It has therefore become essential to provide a robust and consistent image of the area and to ensure that the public realm has been maintained to a high standard. The daily activities of commuting and commerce along with the priorities of residential property owners and large-scale landowners give particular meaning or sense of place to Paddington Basin. Lynch writes that we inform places with our own meaning (1960:92) and it is with expectations of increased land and property value that Paddington Basin is ascribed. During a period of optimism for development in the 1980s, Julian Dean wrote a memo to his colleagues in Westminster suggesting 'We should plan for comprehensive mixed-use development, not on one scheme, but based on ownership, because that gets you to implementation' (Interview 2012). Over the subsequent decades, a high-quality public realm has been realised across Paddington Basin, controlled and maintained, furthering a particular understanding of public space as property in London.

Fourth, public spaces at Paddington Basin are instrumental to the success of the development. We have seen in Paddington Basin, as in the regeneration of Elephant and Castle, how public spaces are provided and often required through Section 106 agreements between developers and the local authority. In Section 106 of the Town and Country Planning Act 1990 these negotiated agreements are termed 'planning obligations' intended to be imposed on the developer as considered 'necessary to make a development acceptable in planning terms' (www.legislation.gov.uk 2015). Developers have increasingly recognised the value of these public spaces to their developments through their provision and on-going control. In addition to acting as an obligation required of the developers, the high-quality public spaces provide an opportunity to frame a narrative of Paddington Basin as an exclusive place, contributing to the image of the developments and thus enhancing their value. The public spaces also provide a buffer around each building through which private security guards can patrol and across which restaurant premises can expand. While visually and spatially permeable, as required by the Canal and River Trust, the privately owned and managed public spaces provide a protective zone from which protestors, homeless people, prostitutes and teenagers—and anyone else deemed out of place—can be excluded. Additionally, through claiming to provide a new, costly and high-quality public realm, the developers can negotiate out of their obligations to build affordable housing or they can benefit from building at a greater volume or height than they may have been permitted to do otherwise. In this sense, Section 106 is co-opted by developers, allowing private developers to gain from the planning controls that are in place to benefit the public.

Finally, I identify that through the development process, public spaces and forms of public are redefined. Operations of cleaning, maintenance, security and events leave the public spaces undifferentiated from a mall without a roof. The multiple entrances and the accessible routes through the masterplan area create the illusion of a public realm comparable with other districts in London where streets, parks and squares are in public ownership. Designed spaces, such as the amphitheatre, suggest democratic forms of public space and street signage mimics those implemented by the City of Westminster. But the ease of which it is possible, for most people, to move uninterrupted through the area across a landscape of high-quality materials with minimal contact with others, frames these public spaces as passive and frictionless environments. The developers have produced a restricted public realm, where participation is circumscribed and the contrast with adjacent public streets, such as the Edgware Road beyond the control of the masterplan and the Paddington BID, is extreme. Kate Beaton described that a junior officer at Westminster Council wrote in a letter to the BID claiming 'that we had created an enclave which did not relate to the City of Westminster' (Interview 2012). In Paddington Basin, I have found what David Grahame Shane terms as an enclave of 'distinct spatial and social orders'. Shane explains that enclaves can be built to 'help to distinguish them from their surroundings' and systems of control and regulations can restrict the 'social and functional order for specific people and uses' (Shane 2005:177). The new public realm, as promised by the developers and as repeated across other places in London, follows specific models of development, redefining how public space is experienced by residents, workers and tourists in London. A new urban condition that favours the priorities of developers and investors has been constructed, as Westminster's Julian Dean concluded: 'It is up to them how they organise [the open spaces]. Because they [Broadgate Estates] are not public, or public land' (Interview 2012).

In this chapter I have analysed the regeneration of the canal basin and rail yards at Paddington: a development that reflects the financial ambitions of investors and developers; a masterplan facilitated by commercially minded public agencies and a laissez-faire local authority; this has resulted in a situation whereby individuals and groups deemed undesirable have been excluded from the reconditioned neighbourhood. Although the developers have attempted to author a narrative of making a distinctive place, *Paddington Waterside* follows a common model for development, including buildings with large floorplates that are attractive for corporate tenants and water views that increase the value of residential property. However, as I question these repeatedly rewritten accounts of place-making I propose that this model for development should be described in terms of the significant public investment required to make it possible and the exclusive public realm that undermines any notion of public life. I have revealed a site of intense speculation based on the value of land and property, a disposal of public assets and the control of people, activities and spaces in an attempt to create a unique place.

Notes

Fieldwork and discussions from this research into Paddington Basin have also been published in: Wall, E. 2022. From cleaning to cleansing: Maintenance as an urban development practice at Paddington Waterside, London. In: T. Viderman et al. (eds) *Unsettled: Urban routines, temporalities, and contestations*. Oxon: Routledge (forthcoming 2022).

1. The term 'Paddington Basin' has been adopted in this book to refer to the waterfront spaces from Little Venice to Paddington Basin, including the open spaces contiguous with it which form the *Paddington Waterside* masterplan (Paddington Basin is described by the Canal and River Trust as the canal basin approximately 500 metres south of Little Venice and it is also the name used by a previous development operation, now called *Merchant Square*). The term 'Paddington' is used in this book to describe a wider area which includes the historic residential areas, commercial streets and the train station.

4

ORNAMENTS AND IMAGES

> 'London is built around its squares and its architecture and its old buildings, and they are beautiful'.
>
> —John Prescott, Former Deputy Prime Minister (Interview 2014)

Trafalgar Square is a space of architectural ornament, curated performances and the production of images. It is the geographical centre of London, the place from which distances to and from the capital are measured, and it is the setting for many historical narratives about London, England and Britain. Trafalgar Square attracts tourists and Londoners alike, as individuals passing on their way to work and crowds drawn to the square's monumental setting. The relentless imaging of the square, through the photographic lenses of cameras, phones and other gadgets, frames many individual experiences. In contrast, the daily use of the square and the regular scheduling of events define particular notions of publicness. From when the square was built in the 1800s to its refashioning in 2003, under the direction of the *World Squares for All* masterplan, the square's visual composition, its ability to open up dramatic views and its setting for the surrounding grand buildings has remained a priority for organisations responsible for the space. Trafalgar Square is a key location employed in the production of World-class images of London. It is the second most photographed landmark in the world (Crandall et al. 2009). The perspective of Trafalgar Square as a highly visible democratic public space also attracts protestors, unions and activists wishing to engage with, or at least to be seen or heard by, powerful political organisations and individuals in the UK. These perceptions of London and its ability to compete to attract tourists and media attention are part of what Aspa Gospodini explains is the prioritisation of 'the production of a prestigious and symbolic urban landscape' (2002:61).

DOI: 10.4324/9781003248187-5

FIGURE 4.1 Team London event staff and Trafalgar Square Heritage Wardens take photos for and with visitors, most often with tourists leaning against the edge of the fountains.

(Photo by author)

In this chapter I explore the third case, the *World Squares for All* masterplan, with a focus on Trafalgar Square redesigned by the government, strongly curated by municipal and commercial actors and continually photographed and disseminated as visual images. I emphasise the remaking of the square during and since its masterplanning, which unfolded between 1996 and 2003, contextualised with descriptions of the decisions that led to its first opening in 1844. I examine how the latest redesign embraced opportunities to frame magnificent views of the square's features and surrounding buildings despite priorities that were claimed to focus on reorganising traffic (Interview with Deputy Prime Minister John Prescott 2014). I consider past memorable public events, such as the demonstrations against the Poll Tax in the 1990s, as informing contemporary schedules of planned events and spontaneous gatherings which have occupied the square. I discuss strategic decisions for renovating the square to increase access for visitors in the context of Richard Sennett's description of how in the nineteenth century public discourses were taken outside of pubs and coffee shops through the design of new public squares (1994:346). These changes to the city, which Sennett claims 'resist the demands of the crowd and privilege the claims of individuals' (1994:369) resonate with the rhythm of events that the Greater London Authority (GLA) permit to occupy the space and the otherwise dominance of individuals, such as tourists, taking photographs of the square. In this chapter I describe the making of the square as a place to find national and international publicity for political struggles and cultural events, as well as for individual visibility in photographs and social media.

This chapter is arranged in three sections that discuss the formation of the square in 1844 and the re-design in 2003. The first section explores associations that are made across the square, from the spatial relationships between an inner square of statues and the surrounding circulatory space, to the association of the square with a history of public events, famous architects and memorable images. In the second section I discuss the ornaments and setting of the square and relations between these spatial forms and the production of images through photography and film making. I examine both the physical actions of taking, posing for and disseminating photographs as well as the formation of individual and collective images of London. In the final section, I consider the square as a place associated with contestations, but which is managed as a public space through control and consensus. It is a public square that is less fought over, but rather a place to be used to gain publicity for national concerns, commercial events and cultural celebrations that are brought to the square under a global gaze.

Scenes of associations

> 'It wasn't, and isn't and was never a Trafalgar Square *only* project, it was a Trafalgar Square *area* project'.
>
> —Simon Townsend, architect involved in the redesign of the square (Interview 2014)

Trafalgar Square is a grand civic square in the centre of London, located above and aligned with Charing Cross. The square is a 12,100m2 traffic island which forms the termination of six streets: the Mall, Pall Mall, Charing Cross Road, the Strand, Northumberland Avenue and Whitehall. Trafalgar Square is a destination, a through-route and a public space. It is a key landmark within central London. Initial intentions for a square at Charing Cross were suggested by architect John Nash as part of his proposal for Regent Street (1814), a proposal set out to remove many existing buildings between Cockspur Street and Pall Mall. As part of the strategic plans to improve movement and communication across London, Trafalgar Square was formed as a traffic island, an intersection with roads meeting around its perimeter. The square was a significant space of movement, an intersection integral to the network of streets and squares that were planned by

Nash and subsequently designed and realised by Charles Barry. An emphasis on movement and the spatial arrangement of central London were also fundamental to arguments for reconfiguring the square at the end of the twentieth century. The networks of streets and squares, the former Deputy Prime Minister John Prescott described, had been 'taken over' by the car (Interview 2014). They formed a system that became what Richard Rogers and Mark Fisher call, a 'rammed roundabout' 'whose centre [was] cut off from surrounding buildings by cross currents of traffic'. It was a place, they claim, which put 'the needs of pedestrians second' (1992:105).

Two contrasting spaces

Since it was first opened as a public space in May 1844, Trafalgar Square has been composed of two distinct but connected areas. The two zones include the central square, which is enclosed by stone retaining walls to the north, east and west while open towards Whitehall, and the surrounding circulation of pavements and roads. The central square includes the ornamentation of two giant fountains, Nelson's Column with its four lions, three statues and busts, the Fourth Plinth, which hosts a roster of contemporary sculpture initiated by the Royal Society of the Arts; a structure that was formerly London's smallest police station; and an entrance to both Charing Cross underground station and the entrance to a subterranean pedestrian tunnel. The tunnel connects the central square to the outer walkways and pedestrian paths that follow the roads, which until the square was remade in 2003 fully encircled the square. Like Regent's Park, which was also planned by Nash in the nineteenth century, this 'wall of traffic', as Sennett describes encircling the park (1994:325), restricted many uses of the square and, in particular, meetings of organised groups. As part of the renovation in 2003, the road at the top of the square that ran in front of the National Gallery was closed to vehicular traffic. John Prescott, the then Deputy Prime Minister, and the minister responsible for planning, described that 'to go out to the gallery there and walk into the square and fight your way through the traffic was absurd' (Interview 2014). The closure of the street to traffic created an upper terrace that brought together the square with the museum as an uninterrupted pedestrian public space. The architect, Norman Foster, who led the masterplan, claims that 'the improvements recreate this major civic space, turning an undignified traffic roundabout into a truly public space' (www.fosterandpartners.com 2014). The planners who worked with Foster claim that along with modifications to the central square, such as the introduction of a new flight of steps in the centre of the terrace, the changes to the traffic flows caused 'levels of pedestrian movement' to increase by thirteen times (www.spacesyntax.com 2014).

The streets, the upper terrace and the central square, a space that since the latest renovation also includes a large public toilet and a café, reveal different rhythms of London. Although directly adjacent, they offer contrasting spatial conditions and what Fran Tonkiss describes as 'different senses of being together in public' (2005:67). Tonkiss identifies the café, the square and the street as three 'ideal-types' of public space. Despite closing the upper terrace to vehicular traffic, formerly the road that extended Pall Mall East across the front of the National Gallery, the terrace can still be considered a street. Tonkiss describes the 'informal encounter' that defines spaces like the upper terrace, where street performers, commuters and tourists intermingle—strangers sharing a communal space where they are 'obliged to accommodate others' (2005:68). Following the transformation of Trafalgar Square, the upper terrace has remained the responsibility of the City of Westminster whose bylaws and regulations differ from those of the GLA who manage the central square below. Regulatory changes enshrined in the 1999 Greater London Authority Act have since redefined the square as two spaces. The 1999 Act describes the laws to be sanctioned by the state to control the activities within the square. The historian, Rodney Mace, writes that from 1848 there have been successive regulations written and rewritten to restrict activities occurring in the square, such as rough-sleeping, public assembly and delivering public speeches (2005:134). Street performers are excluded from the central square through Section 5 of the Greater London Authority Act 1999, which cites that written permission is required to organise a performance, play a musical instrument or solicit money (www.london.gov.uk 2014). The central square is a monumental space provided by the state, under the authority of the Crown, symbolising ideals of public space and offering representations of how the government wish London and Britain to be perceived. However, on the upper terrace, performers have the same claims to the street as commuters and tourists who congregate before entering the National Gallery. Jim Muir, an architect involved with Fourth Plinth Programme, describes that 'you get people down here [in the central square] but this [the upper terrace] is where the action happens throughout the day' (Interview 2013).

Mace sees Trafalgar Square as part of an area in London that 'encloses the symbolic seat of power and the centre of government of the United Kingdom' (2005:11). This is reflected in the *World Squares for All* masterplan, a plan which formed the basis for the remaking of the square in 2003, which extends from Trafalgar Square, past Parliament Square to Millbank. Trafalgar Square marks the intersection of the Mall leading to Buckingham Palace, the Strand which connects to the financial centre of the City of London and Whitehall which looks towards Parliament. The square was built in the mid-1800s on a site previously inhabited by the Crown stables and several taverns, small business and coffee houses; spaces 'representing social exchange', as described by Tonkiss, were replaced by a different form of public space, an open square 'provided' and 'protected' by the state (2005:67). As well as a grand termination of the streets that Nash had built and planned across London, his intention for the square was to enlarge a space of meeting for the city's population. This understanding of Trafalgar Square, as a central meeting space in the nation's capital, as well as a facilitation of traffic, reflects the two elements of the square's composition: Londoners and visitors alike are drawn to the central square for its symbolic and historic associations, its grand views and architecture and for the events that punctuate the year.

An urban renaissance?

In their critique, *The New London* (1992), Rogers and Fisher set out arguments to reconfigure Trafalgar Square and address growing dissatisfaction of the central square, as separated from the rest of London. Their manifesto follows proposals for the capital that Rogers had drawn up and which contributed to an exhibition in 1986 of design speculations at the Royal Academy called *London As It Could Be*. Rogers' architecture practice describes the context for the proposals:

> The election of a Conservative government in Britain, followed by the greatest property boom of the century, led to massive redevelopment in London. Rogers felt great opportunities to improve the capital were being ignored in favour of a piecemeal approach to planning, led by market forces rather than by any consideration of the wider public interest.
>
> (*www.rsh-p.com 2017*)

In drawings and writing Rogers proposed the creation of 'an apron piazza leading down the slope of the square from the National Gallery' which aimed to bring the 'whole space together' (Rogers and Fisher 1992:105). The authors anticipated the election of New Labour in 1997, the formation of a new national government in which Fisher would become Minister for the Arts and for whom Rogers would lead the Urban Task Force. Rogers published a national manifesto for planning with the Urban Task Force, titled *Towards an Urban Renaissance* (1999). The document elevated public space as a solution for what Rogers and Fisher perceived as a 'shabby city playing a less and less culturally central role' (1992:XIV). In addition to urban strategies that confronted their concerns for a decline of public life in the face of laissez-faire free-market development, they described specific spaces, like Trafalgar Square, and proposed solutions.

However, when the architectural proposals for a new Trafalgar Square became a commissioned masterplan it was not Rogers, but rather his former colleague, and architect, Norman Foster who took centre stage. In 1996 Foster was selected to lead a team to define the *World Squares for All* masterplan, which was set out by a consortium of national and metropolitan agencies, to include the areas around Parliament Square, Whitehall, Embankment and Trafalgar Square. The steering group for the masterplan, which included the GLA, English Heritage, the Metropolitan Police, the Parliamentary Works Directive, the Royal Parks Agency, Transport for London, Westminster Abbey and Westminster City Council (www.london.gov.uk 2014), considered the continual flow of traffic around both Parliament and Trafalgar Square as a hindrance to public activity and the movement of pedestrians. As the Deputy Prime Minister John Prescott described 'they are both squares, problemed by traffic, but a different character' (Interview 2014). The

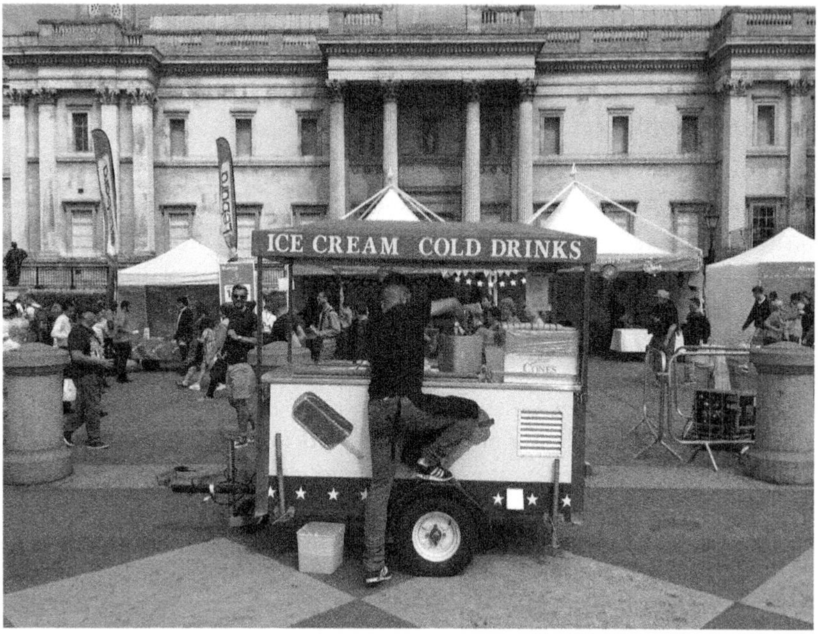

FIGURE 4.2 The redesign of Trafalgar Square closed the upper terrace to vehicular traffic, providing a pedestrian connection to the National Gallery and a place for events as well as busking, performances and protest that are restricted within the main square.

(Photo by author)

masterplan reiterated the proposal cited by Rogers and Fisher to restrict vehicular traffic in front of the National Gallery, which would improve pedestrian connections from Pall Mall and Charing Cross Road towards Whitehall. When this ambition in the masterplan was defined as a project to be further designed and built, commissioning a team, which again included Norman Foster, the design incorporated a new flight of steps aligned with the Gallery and Nelson's Column, new toilet facilities and a café in the under-croft of what had become a new terrace looking down across the Square. The central square was then further connected to the terrace above by two glass elevators. With this deft closure of a vehicular road in front of the National Gallery, Trafalgar Square had expanded to meet the buildings that claim its address and connect more to the activities and public lives of London.

Simon Townsend, an architect who had persuaded Foster to lead his *World Squares for All* masterplan team and who later led the redesign of the square, originally approached Richard Rogers to head his team. Townsend had attended a pre-election debate, sponsored by the *London Evening Standard* newspaper and the Architecture Foundation, where he described the Environment Minister John Gummer's claim that 'two of our greatest civic spaces were just giant [traffic] roundabouts' (Interview 2014). Following the event, the planning department at the City of Westminster formed a 'partnership of interests' across different authorities and responsibilities to commission a masterplan. Townsend 'looked around to who to partner with' to win the commission and, in particular, he described how he looked for 'big-name architects' (Interview 2014). Townsend was aware of the importance of architects, who Sharon Zukin claims are 'chosen on the basis of both their names and their work' (1991:46). Despite the 'huge ego' which Townsend recognises is associated with many 'star architects' (Interview 2013), he was aware that the presence of notable architects like Richard Rogers or Norman Foster within a bidding team would offer a competitive advantage and also legitimise the proposals to remake the square in the way that the high profile of architects John Nash and Charles Barry had done in the 1800s. Zukin describes that 'recently there has been a heightening of both subjective and objective use of individual architects' (1991:46) who inscribe their own 'signature on the landscape' through their designs while also becoming part of the 'landscape of power' (1991:47). Although few Foster 'signature' forms can be found in the redesign of Trafalgar Square, early proposals did include what Townsend describes as 'Fosterillos' (Interview 2014), a conflation of the steel and glass entrances to Foster's Bilbao underground system, which are locally referred to as 'Fosteritos', with another of the architect's buildings, the Armadillo in Glasgow (www.heraldscotland.com 1997). By referring to 'Fosterillos', Townsend is parodying what Zukin calls 'the standardised forms which well-known "superstar architects" . . . move from place to place' (1991:47).

The association of Foster and Rogers with Trafalgar Square, and the connection of the redesign of the square to famous architects, reinforces the cultural value that remaking such a space imbues. Zukin describes the 'economic and cultural value' that is exchanged through projects like Trafalgar Square (1991:46). In cultural terms, Norman Foster and Trafalgar Square bestow one another with value. The importance of one of England's most famous architects (www.theguardian.com 2007) to redesign London's most famous square (www.london.gov.uk 2014) has been reciprocal. Townsend claims that while the square made a loss for Foster and Partners, it has remained one of the 'highest profile from marketing terms' (Interview 2014). Although economically leading the masterplan and being involved in the redesign was less successful, the project enhanced Foster's profile and his association with the design of notable public spaces, masterplans and with the historical distinction of central London.

While Foster and Partners, who led the masterplan, did not lead the redesign of the square they have received continued credit for the project at the expense of Townsend's firm, who led the implementation. Describing the redevelopment of the square, architecture critic Deyan Sudjic incorrectly describes 'Norman Foster's transformed square' (www.theguardian.com 2003) while Foster and Partners' own website quotes the *London Evening Standard*, which opens with: 'Sir Norman Foster's scheme to pedestrianize the square' (www.fosterandpartners.com 2014). As Zukin surmises, 'a larger number of people may be impressed by architects' names than actually know their buildings' (1991:46). Townsend attempts to clarify who authored the Square: 'Unambiguously, Foster and Partners were the lead contracting party for the masterplan from 1996 to 1998'. However, he insists that 'on the implementation the lead was [us]. I mean, [it] was, is and cannot be argued' (Interview 2014). However, it is not the failure to report Townsend as the author which is most problematic; Townsend recognises the value that Foster brings to the team and he expected that he would receive an imbalance of credit. Instead, Townsend is concerned that a different and competing firm has consistently been associated with the project: Townsend describes that, 'every bloody book that comes out and every report, and I don't know whether it is this one or this one, it is bloody Arups [who are credited]' (Interview 2014). Townsend believes that it is due to the pro-bono work on Trafalgar Square's Fourth Plinth Programme that has resulted in Arup, another multi-disciplinary design and engineering firm with whom Townsend regularly competes for work, to be credited.

Powerful associations

There has been a flurry of individuals, corporations and cultural events associating themselves with the square since it was reopened in 2003. These relations can be read through the spatial, visual and social reconfigurations of Trafalgar Square that have been facilitated and encouraged by the GLA. Ken Livingstone, the first elected Mayor and leader of the GLA, embraced the initiative and assumed the chair of the *World Squares for All* masterplan Steering Group and direct responsibility for Trafalgar Square. Part 10 of the 1999 Greater London Authority Act transferred the 'care, control, management and regulation of the Square and its ornaments' from the Secretary of State to the GLA (www.legislation.gov.uk 2014). The new legislation enabled the GLA to promote the square for filming international films, such as, *Captain America* and *Skyfall*, staging cultural events like hosting the announcement of London being awarded the 2012 Olympic Games and commercial activities, such as T-Mobile's mass sing-along. Through new management of the square, global brands were encouraged to use Trafalgar Square and promote themselves through its scenic landscape, and through such image making, associate themselves with London.

In addition to interest for cultural and commercial activities in the square's public protests, demonstrations and events also recognise the significance of Trafalgar Square's setting. Jason Cobb, a trade union event organiser, described the square as 'both iconic and eminently practical' for events up to 25,000 people, providing 'a very clearly defined space' that offers an amphitheatre of steps and a terrace up to the National Gallery (Interview 2013). The square has been designed with reinforced paving to accommodate large vehicles and installations with an infrastructure of water, electricity and communications points that can service events. The surrounding roads offer an ease of access for emergency services and, through its two tube lines, eighteen daytime bus routes and thirty six night buses, it has convenient transportation links. Cobb emphasises the significance of the large public toilets in the square that allow events to occur without the need for temporary facilities (Interview 2013). The paving, steps, plinths and walls that dominate the square are also easily cleaned and, except for a gradual accumulation of chipped paving, are hard-wearing.

Jason Cobb, who has organised trade union rallies in the square described how the square's proximity to Parliament contributes to the impression that 'if you are talking in Trafalgar Square it feels like you are talking to the nation' (Interview 2013). This historic setting that many historic political struggles have appropriated offers a context that other public spaces are unable to provide. These are associations made through newspaper, television and online commentary, a contradiction that Cobb also recognises: 'it is entirely psychological because you talk to the nation through the media' (Interview 2013). The publicity of the events in Trafalgar Square become essentially bound to the presence that the events themselves seek and where the drama and theatricality of the events becomes essential to drawing media gaze.

Through these events, Trafalgar Square becomes a stage set and a theatrical prop. In his essay, 'The Word Itself', J.B. Jackson examines the development of the term 'landscape' and its historic relationship with theatrical scenery. Describing the eighteenth-century use of the term, Jackson claims that it 'had the function of discretely suggesting the location of the action' (1984:4). This definition of landscape is echoed by Michael Rose, a freelance photographer for one of London's media agencies, when he describes how his photographs are 'always trying to relate . . . what is happening in the square and the square itself' (Interview 2013). The events that are attracted to Trafalgar Square are actively vying for presence within global media networks and the features of the square presented within the visual media of film and photography are essential in reinforcing the associations of the event with London and the nation. As a Londoner, Rose admits that he has no knowledge of the history of the square, but in his photos, he is trying to 'get a sense of the square'. He describes his attempts to incorporate 'some of the icons of Trafalgar Square' which are used as 'props' to simultaneously indicate where the event is taking place and to 'hang a story on' (Interview 2013). What Jackson calls 'scenic devices' (1984:5) include the recognisable features of the fountains, facades, buildings, statues, balustrades and monuments of Trafalgar Square. These features must be captured within the photographic frame for the audience to recognise or make associations with the iconic square, with London and the UK.

The most recent visual ornamentation in the square has been the contemporary artwork on the square's Fourth Plinth. Through an eighteen-month cycle, artworks are selected and commissioned to be placed on what, previously, was described as the empty plinth. The Fourth Plinth Programme is led by the GLA and is intended 'to encourage debate about the place and value of public art in the built environment' (www.london.gov.uk 2014). Initiated in 1994 by Prue Leith, the chair of the Royal Society of the Arts, the Fourth Plinth Programme has become a continuous schedule of commissioned sculptures, originally intended for an equestrian statue of William IV. The presence of the sculptures offers a visually arresting contemporary presence within an otherwise heritage landscape.

FIGURE 4.3 Since the redesign of the square more events and festivals have been scheduled within the square, accompanied by additional security, gating and commercial activity.

(Photo by author)

The artworks are often intended to reference a feeling of permanence, the dominance of male military statues as well as the overbearing presence of stone in square. Christy McLean, one of the curators for the GLA describes that: 'Having a temporary sculpture there really enlivens, it reanimates the historic debate about who the figures actually are' (Interview 2013). The sculptures frequently infer the dominance of male statues celebrating war and the lack of everyday people. The first artwork, *Alison Lapper Pregnant* by Marc Quinn, was a three and a half metre high white marble sculpture depicting fellow artist Alison Lapper who was born without arms and with truncated legs. Quinn describes the association that his sculpture has with the statue of Lord Nelson, disabled in battle, although in contrast he sees Nelson's Column as 'the epitome of a phallic male monument' (www.london.gov.uk *2014*). Quinn's concerns reflect those of Mace who describes the statues in Trafalgar Square which depict 'heroic men' while questioning why 'womanly virtues . . . are almost totally absent' (2005:11).

The opportunity that Fourth Plinth artists have to critique the symbolism of the square is also reflected in the *Hahn/Cock* sculpture by Katharina Frisch. The giant electric blue cockerel that was unveiled in 2013 was claimed to stand as an affront to the statue of Lord Nelson situated on the column opposite. Emphasising concerns of the conservative Thorney Island Society the *Daily Mail* claimed that as 'the national symbol of France' the cockerel was 'a reminder of the nation he [Lord Nelson] defeated' (Enoch 2013). Many of the artists have also responded to the setting of the square itself. Before the *Hahn/Cock* sculpture *Powerless Structures Fig. 101*, a four-metre-high sculpture of a young boy riding a rocking horse by artists Michael Elmgreen and Ingar Dragset was resident on the plinth. Elmgreen claims: 'It [the artwork] needs to be in a dialogue' with the square and the other permanent statues (Searle 2012). The interactions between the sculptures, the square, its setting and the thousands of visitors who pass through each day elevate the artwork, as well as the fountains and column as landmarks. Kevin Lynch describes landmarks as 'point references considered to be external to the observer' and he discusses the importance of 'systems of landmarks' as well as the singular qualities of individual structures and buildings (1960:78). The associations that people make between landmarks and their surroundings are fundamental to how they relate to, and navigate, Trafalgar Square.

Placing ornaments and taking photographs

> 'There is never any shortage of photos from anything that we do, from professional, good amateur to rubbishy Instagram snaps'.
>
> —Jason Cobb, union organiser, describing the photographs taken at demonstrations (Interview 2013)

In the five years between which Trafalgar Square was designed and opened to the public, W. H. Fox Talbot invented the camera (1839) and took the first photograph of the square (1844). After inventing the calotype process, through which a transparent negative image could fix a scene and be used to create multiple photographic prints, Fox Talbot's photograph of Trafalgar Square captured an image of London in the 1800s, a time when the forms and perception of public spaces and people in public were changing. Sennett describes that in the mid-nineteenth century the public personality 'split in two' (1977:195). On the one hand, there was the role of 'actors' who increasingly and more proficiently 'expressed themselves actively in public' while at the same time, London saw the establishment of 'spectator[s]', individuals who did not participate, but rather removed themselves from the public in order 'to observe it' (Sennett 1977:195). With the advent of the camera, a mechanical devise was inserted into public spaces that increasingly made distinct, and mediated between, the urban public characters of actors and spectators. These relationships unfolded across the newly made public spaces of London that would begin to replace the coffee houses and taverns as places of a different public life. Reflecting on the reconstruction of Paris in the 1800s, John Urry and Jonas Larsen state: 'What is of central importance is the reconstruction of urban space which permits new ways of seeing and being seen' (2011:160). Trafalgar Square had been planned by Nash to 'afford a magnificent and beautiful termination of the street from Westminster' (Nash 1812 quoted in Mace 2005:31) and for Nash's client, Charles Arbuthnot, the Commissioner of Woods, Forests and Land Revenues, the square would 'throw open to those in Pall Mall a full view of the magnificent portico of [St. Martin's] Church' (Arbuthnot quoted in Mace 2005:37). Over the subsequent two centuries Trafalgar Square has been made and remade as a scene of ornamentation and as a site for the placing of visual props. These transformations have encouraged engagement with 'actors' and 'spectators', and through technological and social transformations they have blurred the separation of these previously distinct roles.

Nash had planned the square as an island, around which carriages would travel. So, when the square was redesigned from 1996–2003, the architects and politicians claimed to address the problematic circulation of roads. Sennett criticises the road systems that Nash had originally planned, which enabled the 'movement of large numbers of individuals in the city' while disabling the 'movement of groups' (1994:324). The extended pedestrian space which opened in 2003 resulted in a heightened presence of actors and spectators. Street performers as statues compete through their motionlessness for tourists' attention while the GLA, custodians of the square since 1999, seek to charge professional photographers and film-makers for recording in the space. The GLA also host image-friendly events, from cultural fairs to political demonstrations, to globally promote the UK's democratic values and traditions, with which London and the UK wish to associate themselves. Mace describes Trafalgar Square as London's 'front room'. Its adornment with overwhelming stone paving, historic facades, sweeping balustrades and oversized fountains has created 'somewhere that will impress the neighbours and overawe the country cousins' (2005:11), while its ornamentation of military statues reminds visitors of the heavy history associated with those who built, own and control this central London public space.

Public images

Trafalgar Square is not merely a scenic urban landscape but a place that has evolved along with photography and image making. It is a place where tourists and Londoners alike converge, documenting the scene and their experience in the square on cameras, mobile phones, tablet computers and camcorders. It is a central London site that has been formed and repeatedly reinvented through global as well as micro events; it has been reorganised with architectural interventions; it has been differentially managed through rewritten bylaws and commercial interests all with an interest in how London is perceived. The square is bound up with a diversity of image making, from film and television to CCTV surveillance—even the police and political protestors record each other during contentious demonstrations. Images of occurrences in Trafalgar Square are disseminated to audiences around the globe, while quintessentially British events (such as Royal ceremonies), to which most people are uninvited, are shared on giant television screens within the square. Such relentless image-making and sharing involves embodied, political and architectural actions, evidenced in the square through the prevalence of visitors taking photographs, the bylaws that facilitate the remaking of the space each day and the architectural landmarks that draw the attention of tourists behind camera lenses.

Sharing of multiple individual images of a place forms what Lynch describes as 'public images', defined as 'common mental pictures carried by large numbers of a city's inhabitants' (1960:7). The many striking views across the square, from the portico of the National Gallery, up towards St. Martin's church and with the fountains in the foreground, frame a 'series of public images' which are photographed again and again each day. But for Lynch, a city image is not merely visual: 'Nearly every sense is in operation, and the image is the composite of them all' (1960:2). Through this conception, notable events, such as the 2009 Camp for Climate Action, which form public images of the square as a place of democracy and public discourse, may not be directly visual or recorded in photographs and may be based on compound media, memories and meanings. Trafalgar Square is frequently adopted as the 'public image' of London as international films, news and literature reinforce a narrow view of London. While the professional image-making of film, television and media events contributes to a place of local and global rhythms, the presence of amateur photography blurs with and dominates the experience of the square. What Urry and Larsen claim to be 'the most important technology for developing and extending the tourist gaze' (2011:154), photography offers a small-scale action that finds presence in individual visits and the gathering of families and crowds.

With its remaking in 2003, Trafalgar Square has become a tourist site which receives and facilitates large numbers of visitors. Through the strategy of the *World Squares for All* masterplan, the square has become a place for which 'systematic, regularised and evaluative procedures'—such as cleaners searching out litter and farriers scaring pigeons with their hawks—have been developed to enhance the square's 'tourism potential' for individual visitors and events (Urry 2001:2). There are several subtle daily rhythms that unfold across the square, such as: commuters crossing the square from about seven o'clock in the morning; opening of the toilets at eight; fountains turned on at eight-thirty; and the changing shifts of Heritage Wardens, cleaners and the falconer. However, rather than a dominance of daily patterns, Trafalgar Square is a place that embodies seasonal rhythms and one-off events. As London warms up during the summer, the square becomes increasingly dominated by the presence of visitors and tourists; as Deputy Prime Minister John Prescott described, 'there are certain places where people come as tourists' (Interview 2014). As the school summer holidays end, the masses of families and English language groups recede and are replaced by large commercial and festive events. Prominently, the erection of the Christmas tree, delivered annually from Norway since 1947, and the turning on of the Christmas lights punctuate the space and the seasonal calendar of traditions, which also includes celebrations for Chinese New Year, St. Patrick's Day, Diwali, Eid, Vaisakhi and St. George's Day. Through its management by the GLA, what Urry describes as an 'omnivorous producing and consuming of place' with 'corresponding visual images' (2001:2) has been facilitated through the hosting of large day-long or weekend events that momentarily unpack their infrastructures in the square, welcome thousands of people and then return the square as they found it.

Beyond the annual events, an eighteen-month cycle of remaking the square is evident through the Fourth Plinth Programme. Jim Muir from the Fourth Plinth Commissioning Group claimed that the temporary artwork on the plinth 'is part of the rhythm of London.' He described how 'there is a piece that comes up every eighteen months and there is a celebration of that and it creates a lot of debate and a lot of nonsense in the press and there is an event . . . and it is part of the season' (Interview 2013). In the lead up to the unveiling of *Hahn/Cock* by Katharina Frisch, the GLA widely advertised the coming sculpture. Posters showing close-up shots of the sculpture tempted commuters on the Tube to follow the unveiling on Twitter. In addition to the commissioned sculpture, which was the central feature of the programme, there were supporting press and VIP breakfasts, unveilings and evening receptions that marked the public presentation. They formed a programme of events that Christy McLean, who leads the Fourth Plinth Programme for the GLA, believed raises the 'global attention' and enhanced the profile of London: 'It is important for the Mayor and it is important for London' (Interview 2013). But like all large, globally visible events, the insertion of new sculptures on the fourth plinth is facilitated by spatial transformations in the square that further influence the visitors' photographic frame.

Small to large events

On days when there are no events, tourists, as well as some regular visitors, take seats around the edge of the square. Sennett recounts that in the Parisian cafés of the 1800s, the spectators sat 'silently watching the crowds go by'; similarly, in Trafalgar Square visitors look into the square 'appearing as scenery, as spectacle' (1994:346). The seats in the square are popular with families, couples having lunch and individual people-watchers. There are other architectural features that also promote sitting, including, the ledge around the two fountains and the three flights of steps. Conversely, actors being watched include groups of families and friends, people taking shortcuts through the square, tourists taking photos, children chasing the pigeons and groups of teenagers who climb onto the enormous bronze lions. On the long straight benches along the edge of the grand civic space, visitors do not face each other to speak but look towards the centre of the square as if 'lost in his or her own thoughts' (Sennett 1994:346). The seats to the east are shaded in the morning and those in the west are shaded later in the day, leading visitors to migrate from one side of the square to the other to avoid or to follow the sun.

The regular daily activities of the square are interrupted during events. On these occasions, whether planned or spontaneous, the scale of gatherings subsumes the presence and opportunity of individuals. Groups spontaneously assemble around issues of public concern and discourse, such as the unregulated celebrations that followed the death of Margaret Thatcher, but mostly they are crowds awaiting scheduled entertainment or presenting in public causes that have secured a licence in advance. On the day before a planned event the square is mobilised 'as spectacle' to both attract and accommodate 'large numbers of visitors' (Urry 2001:7). For popular events, such as the NFL Fan Rally, galvanised steel barriers are unloaded from lorries and arranged at the top of the square by the National Gallery and lower down adjacent to Nelson's Column. Once enclosed, teams of roadies and contractors construct temporary stages and giant televisions in the square beneath Nelson's Column and erect merchandising tents to flank the edge of the central square. Political rallies are prepared in a similar way: part of the square is fenced off, stages are set up and access is controlled by security and event marshals. The physical presence of crowds of up to 40,000 people (www.visitlondon.com 2014) as well as the barriers, stages and tents put in place to accommodate these events, restrict the use of the square for sitting and watching or for posing for photographs. Flows of commuters are impeded or redirected, tourist photographs are obstructed, and the fountains are sometimes turned off to facilitate the events.

While Urry describes the production and packaging of places through global television (2001:2), a process that Trafalgar Square engages, events that occur outside of the square are also projected within it, such as the screening of England's World Cup football games on giant TVs. As well as communicating what is happening in the square to audiences around the globe via televised and photographed events, the installation of large TV screens transforms Trafalgar Square into an outdoor room in which memorable events are shared with others. These events, which have also included the showing of Royal weddings, generate 'intense moments of co-presence' (Urry 2001:5). They realise the square as the nation's 'front room' (Mace 2005:11), either as a formal parlour for watching events or an everyday living space for relaxing with friends. Maurice Roche (2002:224), explains that events stage and project the 'power to transform' what are otherwise 'mundane places' into host sites. While Trafalgar Square has distinct architectural qualities of listed buildings, monuments, statues, floorscapes and trees, recognised as a City of Westminster Conservation Area, Mace comments on the ordinary dullness of Trafalgar Square and the ornamentation around Nelson's column. He believes that neither the square nor the column would 'warrant more than scant mention in the canons of architectural or sculptural history of the period' (2005:14). But reflecting Roche's comment about the relations between events and their settings, union organiser Jason Cobb described that to have an event hosted in the square 'gives you a sense of it becoming an event of some importance and the theatricality and the history of having events' (Interview 2013). This interest in the spectacle, as Roche explains, offers 'possibilities for people to reanimate the sense of their world' through participating in such events or engaging with them from afar (2002:224).

During the summer, on days when it is not raining, some of the first people to reside on the upper terrace are street performers. Often groups of young men, speaking limited and broken English, dress in costumes attached to inventive elevating metal structures. Arriving early in the morning, to guarantee the most lucrative spots, they place trolleys containing their steel structures and costumes on the pavement outside the National Gallery. Between six and seven o'clock in the morning the trolleys are placed amongst the growing flows of commuters who walk between Charing Cross Station and the West End. Once a space is secured, the performers lay on the grass to sleep, or sit on the wall in front of the National Gallery. Working in teams, the performers keep watch over their equipment and the positioning of other competing actors as they anticipate the masses of tourist groups who will visit Trafalgar Square that day. The presence of organised tours of English language students are recognisable from their branded t-shirts and bags; Martin, a language teacher who organises student tours to London, described how Trafalgar Square is one of several expected sites that is 'part of the programme that parents expect' when children visit London (Fieldwork conversation 2013). During the summer, the first groups arrive around ten o'clock in the morning, just before the National Gallery opens, at which time the performers complete their final preparations. As they help each other into their costumes to stand motionless, like the military statues across the central part of the square.

Taking photographs

While the performers are restricted to the upper terrace, in part due to the bylaws that restrict performances in the central square, the tourist groups move effortlessly between the upper pedestrian concourse and the main square below. Jim Muir from the Fourth Plinth Commission emphasises, 'you get these two kinds of lines of people converging here: of course, all the living sculptures and pavement artists and [then] everyone else' (Interview 2013). The individuals and families who arrive early, between seven and nine o'clock in the morning, share the street and the square with businesspeople who rush along the upper terrace or through the square and with the teams of contractors who prepare the square for the day. As the tourists take photographs of and with performers they are encouraged to give money in return, an exchange that Susan Sontag describes is part of an 'aesthetic consumerism' which confirms and enhances their experience (1977:24).

The earliest tourists, who arrive around eight o'clock in the morning, also record their visit on cameras, phones and tablets. Putting down their coffee cups on the edge of the fountain, early tourists 'convert experience into image' to take home or upload as their 'souvenir' (Sontag 1977:9) of Trafalgar Square. Photographs are taken in the same way as postcards are purchased at other tourist sites across London where the visit to the square becomes validated via photography. Michael Rose, a freelance photographer for a large photography syndicate, explained: 'I think the tourists wandering around Trafalgar Square, with their cameras and iPhones, or whatever, they are drawing their moment in the square and they are trying to capture it' (Interview 2013). Rose describes how tourists are 'trying to remember' the square through taking photographs, reflecting the art critic John Berger's ascertain that the photograph preserves the memory of the scene (2013:52). Through taking photographs visitors have their visit to Trafalgar Square 'fixed' as a visual record before moving on to another key London destination.

The roles of actors and spectators, which Sennett recognises emerging around the time when Trafalgar Square was built (1977:195), vary and shift throughout the day. These two groups, as defined by Sennett, are further elaborated by Roland Barthes

(1981) who describes the photographer as the 'operator' and the 'spectator' as the person viewing the photograph (Barthes 1981:9). In Trafalgar Square, however, the presence of people solely taking photographs as photographers is less present. Instead, almost everyone takes, or has the ability through digital cameras and camera phones, to take photographs. Within moments individuals shift from spectators of scenes in Trafalgar Square, to operators of the photographic equipment, then to spectators of the photographs digitally displayed on camera and camera-phone screens. As Urry and Larsen describe: 'It has become ritual to examine the digital-camera screen after a single shot . . . so that the image can be seen properly' (2011:181). The actions of tourists in Trafalgar Square modifies as friends and family members exchange cameras or occasionally ask the Heritage Wardens to take a group photograph. Visitors switch positions from photographer to subject, moving from self-awareness when being photographed to an awareness of their surroundings and each other as photographer. This transference of roles is compounded with the increased popularity of visitors to Trafalgar Square posing for selfies. A way of taking photographs of oneself, which gained popularity through lightweight cameras and inexpensive digital film, the actor and spectator momentarily become one, with individuals and small groups extending out their arm (or monopod) to photograph themselves against the backdrop of the square.

The scene of Trafalgar Square can also be considered a spectacle. Barthes frames the scene, or 'person or thing photographed', as the 'spectrum' (1981:9), a term he adopts due to its relation to the word 'spectacle'. This reframing brings the setting of the square into a socio-spatial relationship with the spectator and what is being watched. Professional photographers, such as Michael Rose, consider the setting and the event in the photographs they take. Rose states: 'we are trying to place what is happening in the environment' of Trafalgar Square. The everyday spectacles of the square may be seen through 'people splashing their feet in the water' (Interview 2013) while the spectacle of what Roche terms 'megaevents' are, in part, invoked by the 'scale and character' of the 'masses of people' (2002:40). Roche considers these as 'dual' spectacles for the people attending the event, as well as those viewing on television or the Internet (2002:189). The 2003 renovation of the square enhanced the opportunity for spectators and spectacles. The new flight of steps created what union organiser Jason Cobb termed a 'natural amphitheatre' on which visitors sit for photographs, to eat their lunch or to watch others in the square below (Interview 2013). The aesthetic considerations of the square and opportunities to take photographs were foregrounded in both the 2003 reconfiguration and the implementation of the Fourth Plinth Project (1999).

The intensity of interest in photography in Trafalgar Square is evident online. Through analysing 35 million images on the website Flickr, a team of researchers from the University of Cornell ranked London and Trafalgar Square as, respectively, the second most photographed city and landmark in the world (Crandall et al. 2009). Three years earlier, a Trafalgar Square Flickr Group was established where the five hundred members shared over four thousand photographs, under the heading: 'Trafalgar Square is the Social Barometer of London' (www.flickr.com 2014). Yahoo!, the media corporation who part-funded the 2009 research and who own Flickr, concluded that the 'standard shot' from the photographs analysed was 'a wide-angle shot of the National Gallery and St. Martin-in-the-Fields church' (www.yahoo.com 2014). While this vista turns its back on the famous Nelson's Column it highlights the richness of recognisable elements that constitute the square. Together with the scenic backdrops of the landmark buildings, the amphitheatre of steps or even the view down Whitehall framing parliament's Big Ben, this abundance of features, Yahoo! claim, are 'all worthy subjects for your lens' (www.yahoo.com 2014).

Managing contestations and maintaining consensus

> 'So, all of the baggage of place and space and management and control issues were effectively, many of them were, codified in Trafalgar Square into a series of quite specific bylaws'.
>
> —Interview with Jason Cobb, union organiser, 2013

Tradition of protest

There is surprisingly little conflict in Trafalgar Square. For a place that the Greater London Authority (GLA) describes as 'a centre of national democracy and protest' that frequently attracts political and religious demonstrations (www.london.gov.uk 2014), there are few visible, physical contestations in the square. The square has a reputation as a site of protest: 'Trafalgar Square is a square of political and civil action, in many ways, you know, and protests' (Interview with Deputy Prime Minister John Prescott 2014). A small handful of highly documented demonstrations that have brought conflict with the police, such as the Bloody Sunday demonstrations (1887) and the student fees protests (2010), have created public images of the square as a site of resistance. These images punctuate a more mundane rhythm of organised and controlled events for which permission must be requested in advance from the GLA, Transport for London (TfL), Metropolitan Police and the City of Westminster. Most days in the square unfold in a managed way by the Heritage Wardens, who oversee the space, its visitors who wonder at the architectural forms, monuments and urban actors. Deviations from the bylaws (Greater London Authority 1999), some of which prohibit putting your feet in the fountains or feeding pigeons, are mainly managed through discussion with and the presence of the Heritage Wardens accompanied by other contractors who are employed to manage the crowds. Even on days when Trafalgar Square hosts demonstrations and political rallies, the tightly controlled nature of the events belie the square's reputation as an object fought over or a setting for public conflict. Instead, it is most often a place to present narrow contestations that manifest in other spheres of life, across London or the UK, brought to this public forum and managed through consensus, planning and agreement.

For rallies and demonstrations to use the square an organisation must complete an online enquiry form, the same process that commercial and promotional events must follow, forcing political and democratic traditions to fit into a general timetable of events. The 2009 Climate Camp, International Pillow Fight day and Pasar Malam organised by the Malaysian Tourist Board join a schedule of managed proceedings that seek a public platform from which their concerns, issues and products can be seen. In addition to this, Jason Cobb, who has organised events in the square, reiterates the need for permission from the different agencies: 'there is a lot of pressure on Trafalgar Square' and as a result it is not always possible to get permission to use it. He explains how stakeholders like TfL and Westminster Council have 'various controls' over the square and emphasises the importance of consulting with them all and planning far in advance. Cobb believes that his trade union has 'a good track record of planning events' which reduces conflict with the authorities. Participants of demonstrations are drawn to the solidarity and belonging that events offer; however, the main objective of protests is to highlight concerns through the presence of thousands of people occupying the highly recognisable Trafalgar Square captured in images, amplified through publicity on television, newspapers and websites (Interview 2013).

Trafalgar Square has become a site for publicity. As a backdrop to be photographed against, a place to demonstrate in or a location to host an event, Trafalgar Square offers exposure to audiences within the square and further afield through multiplying forms of media. For commercial, cultural and political events, Trafalgar Square provides a setting that amplifies the messages that take to its stage. Jason Cobb explains how organised rallies unfold:

> The standard format for a political demonstration is that you start off somewhere, you assemble everyone, and you have a rally of some kind with speakers and perhaps a little entertainment and then the people disperse.
>
> *(Interview 2013)*

However, he describes that his 'objective is to get as much media coverage as possible'. Times have changed from when rallies and demonstrations were featured in newspapers with 'mass ranks of union banners'; instead the 'classic demonstration shot these days' is 'someone photogenic holding an amusing sign' (Interview 2013). Demonstrations focus on Trafalgar Square because of the publicity that the space offers. Eye-catching placards and press-friendly soundbites distil radical causes as well as commercial ambitions, hoping that the messages will be captured by photographers and syndicated to the national media.

Trafalgar Square offers a setting in which public events can be performed rather than a place in which public discourses emerge. Since it was reconfigured in 2003 and new regulations for events were put in place the square has been reinforced as a place to be looked at, indeed: 'places of the gaze rather than scenes of discourse' (Sennett 1994:358). It has become a place to represent oneself in public, or even bring attention to public concerns, rather than forming and contesting new forms of public. Organised demonstrations and rallies bring issues to the square with the hope of gaining publicity for their causes. The formal surroundings of Trafalgar Square, which Mace describes as 'a palpable expression of its host's social, historical and political aspirations' (2005:11), provide a visibly British context against which pressure groups, organisations and campaigns can present their concerns. Rather than encouraging public discourses, the square is a space for presenting previously formulated arguments. This emphasis on presenting political issues reflects priorities for Trafalgar Square in place since 1844, to transform a place of public discourse into a large public space of spectacle. The fountains put in place by the square's architect Charles Barry, which were intended to prevent large gatherings (Mace 2005), have themselves become the subject of the photographic gaze of visitors. The visual arrangement of the square in individual photographs overwhelms the narratives of Trafalgar Square as a place of free speech or democratic discourse. From the 1796 survey by Thomas Chawner, two coffee houses, two inns and an ale house can be identified, all of which were demolished to make way for the square, representing a shift from what Tonkiss classifies as public spaces of 'social exchange' to spaces that are 'monumental' and 'symbolic' (2005:67).

Contracting responsibilities

Since 2000, the setting of Trafalgar Square has been maintained by several teams of private contractors. From cleaning, horticulture, maintenance, stonework and hosting visitors, contractors are continually present: the GLA describe that 'on any given day, there could be many different people on the square' (www.london.gov.uk 2014). These contractors hold in place the appearance of the square, from the flux of large public events and installations to everyday occurrences. The increased presence of contractors who are authorised to represent the GLA, through the Greater London Authority Act 1999, has contrasted with the reduction of what Jane Jacobs (1961:68), in her research of New York, terms 'public characters' who previously occupied the square. Performers and bird-seed vendors have been displaced by the Act, which prohibits unlicensed performers and traders as well as feeding pigeons. The Act has affected who is permitted to interact within the square and what they are allowed to do. Bernard Rayner is one of the public characters who was displaced. Known locally as Bernie, his family had sold bird-seed in the square for over fifty years. But in 2001 he ceased trading immediately after a settlement was reached following a protracted battle with the GLA.

Replacing Rayner has been a successive team of falconers who are contracted to very publicly keep the pigeons away. The contractors employ a farrier with two hawks, who parks a van on the square seven-days-a-week. Inside the temperature controlled van rests one of the birds while the other is put to work dispersing the population of pigeons, which reached about 4,000 under the care of Rayner. The farrier and his hawks have replaced the pigeons as an attraction for tourists: 'They [tourists] come to see the pigeons—but they take photos of the hawk' described the farrier as he patrolled the square (Fieldwork conversation 2013). A team of Heritage Wardens are also in place to enforce the 1999 Greater London Authority Act, which continues to prohibit feeding pigeons. Despite the performers and vendors being replaced by contractors, the presence of the latter does not automatically qualify them as public characters. For Jacobs, a public character is defined by an individual's interest and ability in being public (1961:68). Instead, most of the time, the cleaning, maintenance, catering, security and pest control contractors provide what William Whyte calls 'mayors'. For Whyte, the mayors of public spaces are 'great communication centers' who are 'quick to spot any departure from the normal life of the place' (1988:160). There is a team of Heritage Wardens that patrol the square 24-hours-a-day with the stated aim of helping visitors and providing information. The Heritage Wardens are primarily there to ensure that the Trafalgar Square bylaws are followed (Greater London Authority Act 1999, Section 385 and Local Government Act 1972, Section 236B) while providing an interface with the police.

The Heritage Wardens appear open to speaking with visitors, posing for and taking photographs. They wear red and blue uniforms, often with a fluorescent yellow jacket over the top. They work in shifts throughout the day and night and walk through the square in a casual manner. Rather than patrolling along a consistent route, the Heritage Wardens meander as they speak with visitors and highlight any minor indiscretions. They ask visitors, who are cycling or who are taking photographs with tripods, to stop what they are doing and after wandering further through the square they come back to remind or enforce their earlier requests. Through their presence and verbal reminders, the Heritage Wardens nudge visitors into complying with the regulations. In the summer, this may include explaining the regulations to visitors who take their shoes off to paddle in the fountains or tourists feeding their sandwiches to the pigeons. Both offences are prohibited by the Greater London Authority Act 1999, which is made explicit in the square through signs and the insistence of the Heritage Wardens.

Towards the end of the working day tourists are joined by Londoners meeting friends, perched on the edge of the fountain with picnics and drinks. Sunny weather encourages larger numbers of visitors, and during warm evenings the Heritage Wardens seem overwhelmed in their task of enforcing the GLA's specific bylaws. When there are too many couples with their feet dangling in the fountains, increasing clusters of friends drinking alcohol or young children feeding bread to the pigeons, the role of wardens in enforcing regulations is ineffective. The square, on these days, performs as a large civic open place where both visitors and Londoners gather; the square, as Tonkiss (2005) frames this ideal type of public space, becomes an extension of Londoners' enjoyment of their city, creating a 'site of collective belonging' that '[affords] equal and in principle free access to all users as citizens' (2005:67). On some summer evenings thousands of people fill the square, an aggregation of small groups meeting together in public.

On most days, less visible than the falconer and the Heritage Wardens, is a team of cleaners. One contractor, Christy McLean from the GLA described, is Paul who is 'responsible for the square . . . the cleaning of the square'. McLean claimed that Paul attended every unveiling of the plinth, has a complete knowledge of the other statues in the square and, in contrast to other contractors, adopts the role of a public character: 'he talks to people about the [Fourth Plinth] Programme all the time and he hears peoples' reactions' (Interview 2013). Quietly the team of cleaners circle the square emptying items from the waste bins. Every thirty minutes one of them walks by lifting individual items of waste from the bins with a litter picking device. While other parts of London empty their bins on a less regular basis, this repeated emptying appears to prevent other visitors to the space from rummaging through them. Through the redesign in 2003, Trafalgar Square has not become a place which physically designs-out the presence of what Whyte calls 'undesirables' (1988) although the presence of the Heritage Wardens and actions of the cleaners, supported by the Greater Authority Act 1999 makes it more difficult for people who search through the bins, drink alcohol in public and occasionally sleep on the benches. The central square is a public space that is controlled through being constantly patrolled, cleaned and managed.

Like the waste quietly removed from the bins, traces of past activities and political graffiti are quickly removed while damage to the square is swiftly fixed. Within hours of large celebrations, organised events and political demonstrations, cleaning contractors cordon off parts of the square to remove paint and posters. When Scotland played football against England at Wembley in 2013 a reported 10,000 cans of beer were cleared from the square. In the evening, after most of the party had receded, a team of 20 cleaners worked from 18:30 until 08:30 the following morning, with Westminster City Council proclaiming: 'Our crews worked hard round the clock to get London back to normal for our residents and businesses this morning' (www.bbc.co.uk 2013). The 'normal' condition of Trafalgar Square is inferred as a space without traces from the night before, whether these are the hangovers of celebrations or the political concerns of those who gathered. The ability of the square to return to normal thus accommodates the schedule of events that frequently occupy the space, the tourists who visit and the need to maintain order and formality in the nation's front room.

Many mechanisms of control in Trafalgar Square are facilitated and supported through discrete image making. CCTV surveillance systems employed by landlords and the state, the Metropolitan Police recording protests, and protestors making videos of the police, contribute to a highly documented and overlooked public space. The regulations of the square restrict how image making can be undertaken while the signs in the square make it explicit to visitors that they are being watched through a network of CCTV cameras. Real-time recording of the square, through CCTV surveillance, is enhanced during protests when the Metropolitan Police film and photograph protestors as potential evidence. This recording of protests facilitates the police in maintaining order before and after events, as well as evidencing the use of Trafalgar Square as a place of democracy and protest. Less evident when visiting the square is the coverage of the square in Google Streetview. Google's mapping coverage of cities around the world is predominantly facilitated by car-mounted cameras and aerial photography. However, Trafalgar Square has been systematically recorded by what Google calls Trekkers, backpack-mounted cameras which can be walked through the city. Through this mapping, Google can geo-reference photographs found on the web and understand the precise location from which the photograph was taken and how it relates to the official Google image of Trafalgar Square. This state and corporate image-making in Trafalgar Square entangle with tourists taking photographs, films being shot, architectural drawings and events being broadcast on large screens—reinforcing the significance of visual images in the continual remaking of this highly visible central London public space.

Conclusions

> '1840 is one of those remarkable moments when the world seems to shift and new patterns of relationships are established'.
>
> —Urry, J. and Larsen, J. 2011. *The Tourist Gaze 3.0.* Sage

In this chapter I have connected the symbolic and visual nature of Trafalgar Square. First, I have revealed that Trafalgar Square has been formed and refashioned as an intensely visited and highly imaged civic space. It is a setting that is understood through photographs and films recorded in the square and it is a space experienced in relation to the taking of photographs. As I have demonstrated in this chapter, the building of Trafalgar Square and the advance of photographic techniques developed at the same time in the nineteenth century creating a strong focus for visitors: Urry and Larsen write that 'from 1840 onwards tourism and photography were assembled together and they remake each other in an irreversible and momentous double-helix' (2011:164). Actions of taking photographs, in contrast to engaging with and in public spaces through collective actions and discourses, further extend Sennett's concern for public life in the 1800s and his descriptions of an emergent individualism of a public separated into 'actors' and 'spectators' (Sennett 1977:195). The GLA's renovation and reconfiguration of the square in 2003 has continued a narrative of communication and the movement of visitors through optimised connections and new spaces for taking photographs. The approach of Nash through his 1812 plan, that can be seen as 'privileging individuals pursuing their own concerns in a crowd' (Sennett 1994:329), has been furthered through contemporary digital communications recreating Trafalgar Square as a place for global tourism and a focus of images that are shared around the globe. We can understand Trafalgar Square as a place of spectacle on which visitors to London descend. The flows of communications are no longer only the spatial conduits of Whitehall or Regent's Street, but the projection

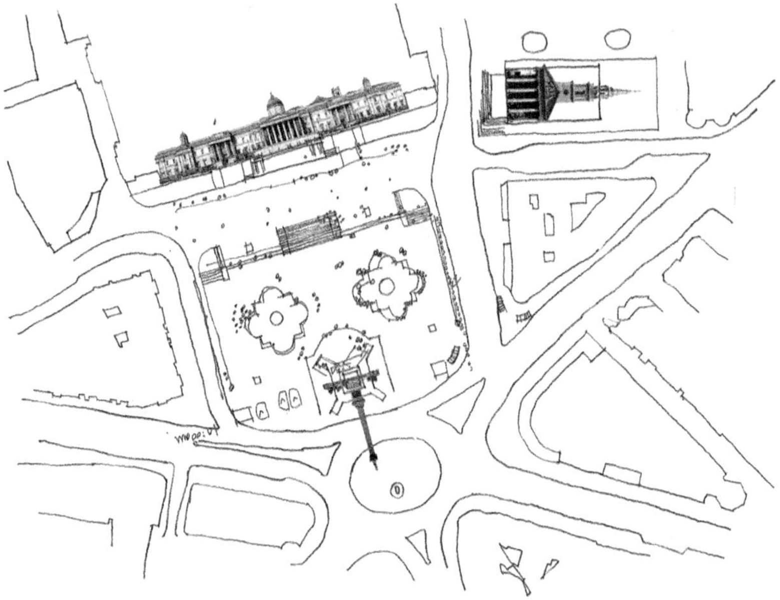

FIGURE 4.4 The contrasting regulations that allow a broader range of public actions on the upper terrace than in the central square are not immediately evident, especially to tourists taking photographs.

(Diagram by author)

of images through Facebook, Instagram and Twitter. Urry and Larsen argue that if photography was not invented then 'contemporary tourist gazes would be wholly different' (2011:186); equally without the transformation of London, through the plans of Nash and his architectural predecessors, neither the gaze nor the image of London would have been framed as they have for the last two centuries.

My second conclusion highlights that there was a significant focus on image making in Trafalgar Square's latest refashioning—and that this focus conceals other regulatory changes. The redesign of the square has further accommodated taking photographs while increased decoration of the square with the Fourth Plinth programme has enhanced the landmark qualities of the space. Gospodini describes cities like London that make 'alterations to the city's image through manipulation of its physical form' (2002:61). The refashioning of Trafalgar Square in 2003, through subtle but significant spatial changes and edited regulations, facilitates this practice of reimaging. Visitors understand the square through eye-level photographic images that include the props of the fountains, the statues and lions in the foreground, brought together by the scenic backdrops of the National Gallery and St. Martin's church. These photographic images ignore the operational separation between the central square, under management of the GLA, and the upper terrace overseen by the City of Westminster. The continuity of materials across the two spaces renders the contrasting regulations invisible. Changes to the regulations in 1999 (Greater London Authority Act) and 2012 (Trafalgar Square Bylaws), along with the spatial reconfigurations, contribute to an ordering of regulations and significantly impacting the way that the space is occupied for public protests and demonstrations.

The third conclusion is that the central square is a space *used* by the public, rather than a public space formed through public actions and discourses. The square is visited by individuals and groups of tourists throughout the year and it is occupied by Londoners and visitors alike for political, commercial and cultural events. Most people who visit the square appear to have no interest in the democratic notions of public space, where ideas of politics are contested, represented and formed. Instead, the scenic qualities of the space enthral visitors and provide opportunities to gain publicity through their presence rather than their actions. Groups wishing to use the space as a site to contest political discourses must follow the same procedures as for commercial and cultural programmes using the square. The need to apply for a permit following the same process as other large events sorts the space by the scale, use and logistics of operation rather than the potential of these actions to define the square as a public space. Similarly, tourists uploading their photographs onto Facebook or protestors vying for online visibility of their concerns are measured in 'views', 'hits' and 'likes'.

Finally, I conclude that Trafalgar Square provides a symbol of democratic traditions rather than a site that supports emerging public discourses. Trade Union Organiser, Jason Cobb, claims:

> It is a public space, it is part of traditional freedom of expression that we have Speaker's Corner in Hyde Park, we have Hyde Park itself and we have Trafalgar Square. And it is part of the British democratic decisions to allow these spaces for dissent.
>
> *(Interview 2013)*

But there appears to be no evidence, from those who have formed the square through spatial and regulatory design, that the square is a place where ideas of democracy can be debated or freely expressed. The square has increasingly come to represent the values that London and Britain wish to be associated: the GLA recognises that 'Trafalgar Square has been seen as a centre of national democracy and protest . . . The Mayor supports this democratic tradition, and gives access to the square for such causes' (www.london.gov.uk 2014). The use of the term 'seen' rather than 'is' reveals the difference between the perception of the square being associated with democracy and the restrictions that have always been in place. That the Mayor 'gives access' to Trafalgar Square for such democratic traditions also conflicts with Don Mitchell's assertion that public space must be 'taken' and made public through public actions (2003:142). From the fountains that Mace claims were intended to restrict large gatherings (2005:88) to the bylaws that instead of prohibiting most activities requires that permission be requested, negotiation is constantly underway between allowing access and defining how Trafalgar Square should be seen.

We have seen through this transformation, a civic square that has been designed to open-up views from surrounding streets and that has become increasingly the subject of personal photographs. This is a place with which many designers, politicians, protestors, performers and visitors strive to be associated. I have highlighted that the image and the setting of Trafalgar Square provide a focus for generating publicity for individuals, brands and issues of concern. The visual cues of the square, such as the statues, fountains and surrounding buildings, provide points of reference for identifying the location of such activities. I have shown that visual priorities are part of the square as a site of public demonstration and protest. In contrast to the masterplanning at Elephant and Castle and development of Paddington Basin, the redesign of Trafalgar Square was initiated and led by public agencies and government departments. However, I have revealed that in common with the other sites, private security and contractors are also employed to enforce new regulations written as part of the physical transformation.

5

APPROACHES TO PUBLIC SPACES

We have found in the preceding three chapters that Elephant and Castle Market, Paddington Basin and Trafalgar Square were repeatedly remade through large-scale transformations, programmed events and daily routines. Across these areas of London there were varying spatial conditions and conflicting practices of making public spaces that can be read through the contrasting ideals and ambitions of the people, organisations and institutions who have produced them. The chapter is structured around three conceptions: ambitions to realise public spaces as *spatial forms*, *visual images* and *social interactions*. First, as *spatial forms*, I identify public spaces composed by urban designers, politicians and developers that are described during planning as brownfield sites and building parcels and redesigned as architectural typologies of squares, plazas and streets. These spatial forms define properties traded for profit and rights of property ownership frequently asserted to challenge access and use. Second, as *visual images*, I describe material public spaces as part of political and economic ambitions, facilitated by architectural renderings, disseminated through various media and retold in filmmaking. I recognise that visitors also defined their relationships with these places through posing for, taking and sharing photographs. These actions frequently conflicted with multiple regulations to control photography, overlapping with the third concept, whereby I recognise public spaces made, remade and contested through *social interactions* within the masterplanned sites. Public spaces were constituted through embodied occupations and physical transformations that resulted from events, markets, gatherings and chance encounters that remade all three places each day.

Spatial forms

The emphasis on economic priorities that local authorities place on urban development and the need for financial profit on the part of commercial developers

DOI: 10.4324/9781003248187-6

results in public spaces that are designed for their *spatial forms*—with bold architectural and visual qualities—but with less concern for public interactions. Whether the masterplanned development was a partnership between a local authority and developers, developer-led, or state-initiated, all three processes reveal strong economic priorities for transforming areas of London. It became evident across all three redevelopments that, and as Ali Madanipour (1996:109) describes, public spaces follow the values and priorities of the developers, politicians and community organisations who inform them. Despite political differences between Southwark and Westminster, the two local councils embraced neighbourhood-scale projects that were reliant on developer involvement. These relationships with developers, which facilitated the *Elephant and Castle Regeneration* and *Paddington Waterside* masterplans, brought millions of pounds of financial contributions into public infrastructures and generated significant profits for developers and investors. The local authorities were dependent on Section 106 contributions, and for this reason they took the risk to make initial investments to enable redevelopment. Southwark were responsible for decanting the Heygate Estate of its residents, while Westminster facilitated the hand-over of land from public agencies. As such, the planners and councillors had a lot to lose if the spatial transformations did not go ahead. In contrast to the profits gained by developers, at Elephant and Castle and Paddington Waterside, the masterplan that realised Trafalgar Square engaged in a global contest. The Greater London Authority (GLA) focused on facilitating the movement of visitors into the square through pedestrianisation. They also created conditions for investment, locally for the GLA through media events, and nationally as these events enhanced the image of London, England and the UK.

As economic priorities came to bear on the developments, outdated spaces were architecturally refashioned, and marginalised individuals including young people, homeless people, sex workers, buskers and traders on the low-cost market became less welcome. At Elephant and Castle, planners questioned the suitability of traders and managers on the old market to operate in the new market square. Similar exclusions of certain people have been evident at Paddington Basin and at Trafalgar Square, through several evictions during the processes of redevelopment, the rewriting of legislation that impacts public spaces and the enforcement of private interests. One of the key objectives of the GLA's *Manifesto for Public Space* (2009) was for a 'prosperous city that can compete nationally and internationally attracting and fostering businesses that bring jobs and growth' (2009:2). However, this redevelopment of public spaces that provide a way of making London more attractive ignores existing low-income jobs, local economies and livelihoods that are unable to continue in these new spaces that operate under different regulations. Activities and uses within public spaces that threaten the investments of local and metropolitan governments, as well as the potential profit of private interests, were therefore found to be restricted.

Masterplanning processes were facilitated by the continued talking down, by politicians and their consultants, of the material spatial forms of existing neighbourhoods. Images of failed buildings and neglected spaces, that are frequently associated in the media with undesirable or criminal activities, were highlighted through employing selected photographs, scripted soundbites and narratives of decline. Derisory representations emphasised the failures of historic social and spatial infrastructures, such as: congested, polluting and dangerous transportation systems (Elephant and Castle Regeneration and Trafalgar Square); dilapidated, unsightly and failed architectural forms (Elephant and Castle); and abandoned, vacant spaces offering redevelopment opportunities (Paddington Basin). At times, criticism of the areas by planners and politicians was left unquestioned in news articles. In contrast, accounts of residents who describe low levels of crime in the area when the Heygate Estate was full of residents (www.heygatewashome.org 2015) and the ambitions of the previous owners of the shopping centre, who presented plans for renovation rather than demolition, were less reported on websites and in newspapers. Other negative messages were perpetuated by the blockbuster dystopian film *Attack the Block* (2011), whose filmmakers were permitted by Southwark Council to film scenes within the Heygate Estate.

The talking down of all three sites has provided a foundation for new architectural visions to be proposed by design consultants that legitimise the ambitions of politicians and developers. By undermining the image of an area of London, vested interests were able to, as one of the residents of Elephant and Castle described, 'stimulate the idea of an obsolete place, a failing place, a place that is full of crime, and noisy, and dirty'. Negative rhetoric was employed to establish a confidence that 'we [the council, developers and consultants] can change all this' (Interview 2012). Negative imagery was also employed by decision makers to argue for demolition through comprehensive redevelopment rather than small-scale local initiatives that were more complex, difficult to visualise and likely less profitable. Such talking down across timeframes of development, to months and even years of threatened disruption, left residents and workers uncertain of their futures.

Visual images

Visual images were rendered to promote the proposed transformations. Another advantage of talking down existing urban conditions is that negative narratives establish opportunities for alternative visions to be proposed. Visual and written proposals were created by urban designers to communicate and fulfil political ambitions and financial goals for the urban redevelopments. Design proposals were necessary to attract investment, to persuade stakeholders and to entice the media. Artistic styles of representing the physical and social transformations were carefully selected: from Foster and Partners' hand sketches, displaying social scenes of a new craft market at Elephant and Castle and a jostling Trafalgar Square, to polished computer renders that reassured investors in Paddington Basin's buildings of the high-quality finishes of their spatial product. Artistic images, which are often distinguished more by what they conceal, were disseminated through websites, newspapers, television and marketing materials. Furthermore, proliferation of visitor photographs taken in public spaces and shared on social media contributed to the public images of each site, influencing how expectations of the newly designed places were formed.

During masterplanning, the creation of new public spaces defined as architectural forms, and practices of recording and sharing scenes of spaces as visual images, were often conflated—as J.B. Jackson describes: 'first [landscape] meant a picture of the view, then the view itself' (1984:3). Renderings of the three cases were commissioned and produced, reflecting the economic priorities of the masterplanning processes. As proposed new public spaces were communicated in architectural drawings they were rendered as landscapes where 'landscape is bound into the market-place and is available only at a price' (Corner 1999:157). In the *Elephant and Castle Regeneration* and *Paddington Waterside* masterplans the imaging of proposed public spaces, which were part of the public access that the local authority required the developer to provide, prioritised the enhancement of the value of the developer's properties. In contrast, in the architectural drawings of Trafalgar Square during its redevelopment and in its imaging through photographs, notions of heritage and culture were brought to the foreground through historic and cultural ornaments. As if critically reflecting Trafalgar Square, James Corner writes: 'It is through styling (design), of course, that one imbues the landscape with allusions to regional and cultural identity' (1999:157).

Each masterplan was also spatially configured with an emphasis on visual composition, the control of views and opportunities to take photographs. This approach to making public spaces as architectural projects reflects the GLA's first objective in their *Manifesto for Public Space*: the GLA describes 'a beautiful city where the spaces between the buildings can inspire, excite and delight visitors and Londoners alike' (2009:2). These are public spaces objectified as landscapes, establishing grand stately vistas, foregrounding developments and placing activities to attract custom. Extending his critique of landscape images to physical landscapes, Corner claims that 'the scene itself displaces viewers, keeps them at a safe and uninvolved distance, and this presents the landscape as little more than an aesthetic object of attention' (1999:156). The structuring of public space through 'scenes', inextricably bound to culturally or financially significant buildings and punctuated by landmark water features and artworks, was present in the plaza, the canal-side and the square. It is an approach criticised by Don Mitchell (2003:186), where public spaces are composed as landscapes in which priorities for order and control overshadow marginal or unappealing daily activities. This structuring of public spaces was coupled with extended visual control as all three sites displayed signage reminding the visiting public that they are under surveillance by closed-circuit television. Restrictions on public interactions limited the participation of some individuals and organisations who wished to engage with the spaces as public sites. In this way, narrow terms of public space as places of safety and spectacle were formed through the overlapping intentions of the developers, the local authorities and the GLA.

Social interactions

Complicated ownerships belie the visual and spatial clarity expressed in the masterplans of Elephant and Castle Market, Paddington Basin and Trafalgar Square—and these come to bear on the social interactions of each site. Across all cases it was necessary to navigate disorganised and occasionally contested conditions of ownership and management. At Elephant and Castle, a patchwork of land parcels that existed since the post-war reconstruction of the area confused the council and shopping centre owners; at Paddington Basin a lengthy court battle over management responsibilities and service charges brought opposing developers and leaseholders to a legally binding situation; and, at Trafalgar Square the different ownerships of the upper terrace and the main square allowed buskers, vendors and cyclists to circumvent the regulations put in place in 1999 by the GLA. Mitchell describes that this 'illusion of control is one aspect of making over a city as landscape' (1997:325). Architectural proposals and their associated visualisations offer confident and definite solutions to the problems affirmed through the initial talking down of the three areas. The apparent 'transparency' of public spaces, as they are reinforced through a certainty of ownership, management, use and architectural drawings, gives false confidence to clients, whether these are the local authorities, developers or people who may use these spaces. The privatisation of public spaces, through their shifting ownership and management from the state to private interests, has for several decades been central to public space debates. However, the uncertainty of ownership and the complexity of management across each site suggest that the ownership of public space may not in itself be the main problem. Instead, it is the power that ownership affords that creates unequal opportunities for different people and organisations to engage with and in public spaces.

In addition to the enormous impact on individuals, families and businesses from the decanting of residents from social housing, closing local schools, running down commercial leases, evicting long-established vendors and closing off historic rights-of-way—threats to public spaces through ongoing control were evident in all three cases. As the developments were built regulations and bylaws for each site were rewritten. While developers increasingly recognise the benefits that open spaces offer their developments, holding onto the control of how they were maintained into the future is also important. At Paddington Basin the BIDs have provided a mechanism for extending control beyond the timeframe and spatial bounds of the masterplan. New regulations, written by the developers and enforced by security guards at Paddington Basin and passed by parliament and enforced by Heritage Wardens at Trafalgar Square, point to the need for spatial transformations to be accompanied by legislation that reinforces regulations that restrict use of public spaces.

The architectural masterplans of all three cases were the primary approaches to the redevelopment processes. The masterplans were attempts, as Corner describes (1999:156), at a 'total vision', not just to represent but also to 'control and condition'. But the drawings accompanying the *Elephant and Castle Regeneration*, *Paddington*

Waterside and *World Squares for All* masterplans were merely visual representations of more comprehensive development instruments, documents, guidelines and agreements. The formation of the BID at Paddington and the new regulations at Trafalgar Square are testament to the expansive roles of the masterplans. However, as with the messy ownership of these sites, the drawings of the masterplans were frequently contradicted by more unstable and unpredictable interrelations of making public spaces. Corner claims that 'the erring realities of life contaminate the purity of any dominant masterplan' (1999:157): realities found in the errors of planning officials in releasing details of the developer agreements at Elephant and Castle, the incursions of undesirable users at Paddington Basin and the Heritage Wardens being overwhelmed by visitors cooling themselves in Trafalgar Square's fountains.

Despite the control of redesigning public spaces being negotiated at high levels of local authorities, I found that individuals and organisations with power assert their presence in many other processes (from masterplanning to operations and from access to use) of making public spaces. In so doing, scalar relations of power can be read in each case. The market managers at Elephant and Castle decide who can rent a stall and dictate what they were permitted to sell. The Heritage Wardens at Trafalgar Square were empowered by the GLA Act 1999 to remove people from the space. Consultants asserted themselves during masterplanning in attempt to define programme, activities and uses. Organised events as well as large spontaneous gatherings occupied public spaces, restricting their use by others. In all three cases, access, exchange and interactions were carefully negotiated. At times, conflict occurred between security guards or police and people deemed to have transgressed the formal regulations of the spaces. Contestations emerged when enforcement was inconsistent at the privately operated Elephant and Castle Market and Paddington Basin, or at Trafalgar Square when the scale of the gatherings and the number of visitors was impossible to manage or police.

However, except at Trafalgar Square, there were few moments when the public who used these open spaces asserted their presence and concerns to the large-scale decision makers, the councils, the developers and the investment funds. On most days and for most people at Elephant and Castle Market and Paddington Basin, the terms of private ownership were experienced as benign, as rights of access across public spaces were maintained by teams of contractors. However, the same private rights were aggressively enforced when surprising, conflicting or undesirable uses emerged. Nevertheless, on the margins of all three sites, slivers of space were found where the rules that have been written to accompany the masterplanned developments were ineffective: at Elephant and Castle, men distributed religious leaflets on the narrow pavements above the market, bypassing the market fees; at Paddington Basin students identified public rights-of-way on the canal towpath on which they could remain unchallenged; and, in Trafalgar Square, unlicensed protests occupied the upper terrace.

The large-scale nature of the masterplanned developments, and the distance and hierarchy from which decisions have been made, limited the ability of planners, commercial interests and consultants leading masterplanning projects to consider these smaller-scale spaces, temporal activities or daily lives. Elephant and Castle Market, Paddington Basin and Trafalgar Square were locations where essential small-scale activities took place: they were sites of people meeting friends, sharing spaces with strangers and exchanging goods despite the marginal spaces they were afforded. I found in all three cases that the ability of decision makers to identify the value of small-scale activities, deteriorated spaces or the needs of individuals, was overshadowed by greater concerns for economic development and profit. Madanipour writes that 'marginal public spaces . . . are not on the list of priorities' when local authorities are considering redevelopment (2010:113). These conditions were exacerbated in the context of unbalanced agreements between developers and local authorities at Elephant and Castle and Paddington Basin. The provision of public goods by the developers is a planning obligation, but these commitments by the developer are only agreements in exchange for other benefits. At times, negotiations and the resultant agreements were concealed, whether in deals between Southwark Council and Lend Lease that local resident Paul felt was 'kept in a very narrow place' (Interview 2012) or where former teacher Judy Peterson questioned the closing of the school in which she worked to make way for the Paddington Basin development (Interview 2013).

Conclusions

I recognise from the findings that overlapping concerns from issues of large-scale developments to economic priorities and from talking down existing areas to promoting future proposals, primarily relate to the masterplanning practices of urban redevelopment. Repeated expressions of economic priorities and on-going control in the production of public spaces established contexts for, and followed processes of, large-scale development. These commonalities raise questions about the development of public space in London through masterplanned redevelopment—issues that are further discussed in the following chapter. In contrast to the masterplans, smaller-scale activities and uses of public spaces highlight the distinctive qualities of such public spaces. The daily exchanges observed in Elephant and Castle Market, the swathes of commuters passing through Paddington Basin and the routine use of Trafalgar Square make these three London geographies exceptional.

While the three processes of masterplanning contrasted (between local, metropolitan and national government and private developers), there were common concerns about the relation of large-scale practices to small-scale public lives of each site. I have shown that plans for comprehensive urban regeneration dominated the spatial forms as well as the everyday lives of the people who live, work, and pass through all three sites. We have seen that threats were exacerbated when small commercial activities that can define the daily presence of the areas no longer align with the images desired by the council, GLA or developers. Even when they were not ultimately displaced, individuals and organisations within these spaces operated under the threat of redevelopment. This affected not just marginalised individuals but also landowners and businesses whose operations could impede the progress of the masterplanned process.

Despite the dominance of economic imperatives and large-scale ambitions, people continued to occupy and use the deteriorated conditions of the moat at Elephant and Castle, enter and pass through the intensely controlled spaces of Paddington Basin and were attracted to the highly programmed spaces of Trafalgar Square. Social interactions, of varying scales and both public and private in character, redefined these public spaces amongst the larger cycles of architectural change. However, as I conclude in the following chapter, opportunities to inform the making of these public spaces remain unevenly distributed and aggressively challenged by individuals and organisations with power across these three London sites.

CONCLUSIONS

I opened the book with a quote from Simon Gregory, a historian local to Elephant and Castle, whom I interviewed in 2012. In our initial email conversation, Simon wrote: 'Any major redevelopment naturally produces a fair amount of controversy, especially if the world detects any unfairness'. Reflecting on the imminent transformation and the large-scale privatisation of Elephant and Castle through the Lend Lease masterplan and a recent exhibition set up by an opposition group, Simon touched on the issue of 'fairness'. I had not anticipated that Simon's concern would remain present throughout the duration of the research as I explored unbalanced relations between masterplan processes, urban spaces and public lives at Elephant and Castle Market, Paddington Basin and Trafalgar Square. The main aim of my research was to explore public spaces through sociological and architectural perspectives to understand contested relations between diverse ways that public spaces are made. As I analysed the interview transcripts, field notes, documents and images, complex and contradictory issues arose to reveal unequal processes of spatial and social transformation, many of which have been articulated by state actors.

Interviews with local residents, commuters, workers, security guards, developers, planners and politicians have described nuanced narratives of how the three London sites are made. These accounts of public spaces designed, built and occupied have been both reinforced and questioned through observation and document surveys. All three cases of masterplanning include public space sites of daily routines where activities of commuting to work and relaxing over lunch take place unhindered. However, it has been through understanding relations between people and spaces, of market stalls erected, clusters of smokers, school kids hanging out after school, men playing checkers—and the potential displacement of these activities—that has brought to the foreground issues of unfairness. The research has revealed spatial, visual and social ambitions that individuals and

DOI: 10.4324/9781003248187-7

organisations have expressed in the remaking of Elephant and Castle Market, Paddington Basin and Trafalgar Square: approaches that have opened up as well as restricted opportunities for others through planned physical actions within these sites and through larger-scale processes of planning and development.

The final conclusions are structured in two parts. First, I draw from the three London cases to present conclusions that critically discuss the imposition of masterplans, the importance of visual images and the power of large-scale strategies. Second, I describe universal considerations that relate to how public spaces can be reconceived in London and in other cities around the world. I discuss the potential of conceiving public spaces as situated socio-spatial processes, and how such an approach could support new forms of, and engagements with, how public spaces are continually produced.

London concerns

Imposing masterplans

First, we can recognise that the masterplans of *Elephant and Castle Regeneration*, *Paddington Waterside* and *World Squares for All* were not comprehensive forms of development, but compromised and piecemeal ways of producing public spaces. As the three redevelopments moved forward, the masterplans were repeatedly amended and sometimes entirely redrawn. Some of the reasons for this incremental progress include: unstable global economic conditions impacting developments and dictating the pace and form of the Elephant and Castle regeneration; designs for residential accommodation superseding plans for office space at Paddington Basin as London's housing market surged; and, the failure of the *World Squares for All* masterplan to progress beyond the first phase as relations of power shifted when Mayor Livingstone was succeeded by Mayor Johnson. At Elephant and Castle and Paddington Basin, progress was also impacted by the reduced role of local authorities following austerity-focused government agendas. Masterplans provide urban designers, developers and politicians with useful tools for communicating total visions for redevelopments. Masterplan documents, such as design guidelines, codes and drawings, reassure stakeholders and encourage investors. The imposing top-down perspective of bold architectural forms, presented in masterplan drawings, follows a tradition of architectural representations that can also offer assurances of order and clarity. Despite definitive plans, in each case the large masterplans progressed sporadically and as their public spaces were realised, the orderly scenes promised in the drawings needed to be reinforced through additional regulations, security and management.

The totalising visions of the masterplans concealed confusing ownership and management boundaries: ownerships that needed to be resolved at Elephant and Castle to progress development; ambiguous management agreements that were exacerbated at Paddington Basin as development parcels were traded between developers; and, a division of responsibilities between the Greater London

Authority (GLA) and the City of Westminster that oversee two differentially regulated areas of Trafalgar Square. The lines of ownership are imperceptible across the redesigned areas and they have no effect on most visitors, demonstrating that for most of the time, terms of ownership have negligible importance for the routine functioning of these public spaces. However, I found that the boundaries became points of contestation as market traders were poorly advised by the council (Elephant and Castle Market), as protestors became aware of the subtleties of ownership (Paddington Basin) and as street performers clustered in some areas while avoiding others (Trafalgar Square). The potential for conflict was heightened as many of the security guards and other privately contracted operators were unaware of the existence or the exact location of these legal boundaries. While these boundaries of contestation underline the contradictions of the masterplans and highlight the difficulty for local authorities to consolidate ownerships, businesses and lives are severely impacted as order is reinforced through the daily operations of the architectural public spaces.

During masterplanning, public and private control was brought to bear on the land parcels of the masterplan areas through new regulations, security teams and management. As each site has been transformed through masterplan processes, regulations that govern the spaces have also been rewritten. Ali Madanipour recognises that 'In the processes of urban change, the conditions of accessibility are subject to change, hence changing the nature of public space' (2010:8). Urban redevelopments create additional rules enforced by new structures of management: as I revealed, the central public spaces of Elephant and Castle, Paddington Basin and Trafalgar Square were all managed by commercial contractors. The partnership between Southwark Council and Lend Lease at Elephant and Castle left swathes of new public spaces under the control of private developers. At Paddington Basin, the developers further extended their influence over the public realm by maintaining and patrolling the public towpaths that continue to be owned by the Canal and River Trust while also founding and part-financing the Business Improvement District (BID), which spreads beyond the masterplan area. At Trafalgar Square private contractors were embraced to fulfil the operations of cleaning and security formerly undertaken by the state. Whether prescribed by government legislation or by private developer regulations, certain public activities, which could lead to undesirable images of the areas, were pushed to the margins of the sites. Despite the rules being enforced unevenly, the image-focused priorities of those who directed the spaces resulted in central and visible areas of the public spaces being highly controlled. This left the margins of narrow pavements and towpaths, where it is difficult to gather and where activities are less visible, to be used for unlicensed gatherings, protests, busking and vending.

The piecemeal progress of the masterplans of *Elephant and Castle Regeneration* and *Paddington Waterside* significantly benefitted the developers who were able to renegotiate densities of development, land-uses and development contributions. The prominence and size of the masterplan developments meant that there was a lot to be gained from investing in the projects, both economically and politically: banks funding private developers at Elephant and Castle and Paddington Basin; architects at Trafalgar Square associating themselves with the prestige of the project despite low professional fees; or the Mayor whose reputation is bound up with the success of these projects through their ability to generate positive images of London and attract investment. The promise from developers for comprehensive transformations of their boroughs commanded the attention of local authorities and central governments who responded by improving transport infrastructures, facilitating planning permissions, transferring formerly public assets into private hands, renegotiating planning obligations and the handing over of future control of large areas of the city. However, the piecemeal progress of the masterplans opened up repeated opportunities for developers at Elephant and Castle and Paddington Basin to renegotiate items, such as percentages of social housing required in residential developments, or to change the direction of the project. Relations between the local governments and the large-scale developers became more unbalanced through the redevelopment processes, even though they both had a lot to lose. The need for Southwark Council and the City of Westminster to compete for investment was a financial necessity for them to fulfil their obligations to maintain other public infrastructures of road and rail networks. However, this prioritisation of financial gain from large-scale planning projects appears to have undermined other responsibilities of the local authorities, such as support for public housing and schools that were both removed to facilitate redevelopment.

Visual appetites

I also conclude that the architectural forms of the public spaces proposed in the three masterplans highlight priorities for visual images that represent both the developments and London more generally. The production of visual images realises commercial interests in marketing specific developments, the images fulfil political ambitions to promote investment into London and they satisfy personal interests in visually recording and sharing experiences through photographs. These visual priorities require investment into the spatial production of public spaces—redevelopment of London that often eclipses the publicness of the sites. We have seen that all three places were spatially reconfigured with an emphasis on their visual composition, including the strategic placing of the low-cost market at Elephant and Castle (and conversely, the same image of the market has provided a rationale for its removal), the views of the canal at Paddington Basin and the vistas along Whitehall towards Trafalgar Square. This approach in prioritising public spaces as scenic landscapes, as a way of seeing the city, enforces what Don Mitchell describes as a 'politics of aesthetics' (1997:326). He explains that the public spaces of cities are replaced by landscapes, in order 'to substitute the visual for the (often uncomfortable and troublesome) heterogeneous interactions of urban life' (1997:329). These landscapes are commissioned by developers, public agencies and governments for their exacting spatial arrangements and focus on high-quality materials.

Architectural forms of strong visual quality have therefore become necessary to fulfil such appetites for images. New forms of public spaces, surrounded by distinctive buildings, offer settings for the generation of enticing images by professional photographers syndicated for the news media and visitors recording their experience to be shared on social media. Southwark Council planner, Tom Appleby, described 'one of the central issues' that he had to overcome at Elephant and Castle as being 'the quality issue' (Interview 2012). He explained that how people think of the area 'pretty much comes down to how we perceive the public realm'—a situation that leads to an increased emphasis on the quality of materials, spaces and forms during redevelopment. Kate Beaton, the leader of the Paddington BID explained, just as Tom Appleby described the situation at Elephant and Castle: 'The [Westminster] council is divesting itself . . . The council could not maintain [the public spaces] to the level that people are currently paying service charge' (Interview 2013). Such an emphasis on quality reinforces a perspective of public spaces as finished and curated products rather than as sites that can be lived and redefined through their use and occupation.

The employment of visual images to render and sell the masterplanned transformations underlines a visual approach to public spaces built as architectural forms. As occurs during online image searches of places in London, scenes of design proposals intermix with historic images and a diversity of visitor photographs to create 'public images' of places (Lynch 1960:46). From Kevin Lynch's research into how urban areas are perceived, he writes that 'there seems to be a public image of any given city which is the overlap of many individual images' (1960:46). Public images of the three cases were formed from first-hand experiences combined with images of these spaces from the Internet, film and television media. The importance afforded to these public images, in continuing to attract visitors, tenants and investors, results in struggles over the authorship and content of which images are created and shared. Once the visual qualities of public spaces are given precedence, Mitchell argues, those who own public space exert their 'control over the relations within it' (1997:323). Restrictions were in place in all three cases for taking photographs. While not described in explicit terms, activities that could result in less favourable images of the public spaces associated with the developments were forbidden. The need to seek permission or pay fees in advance of taking photographs lay in contrast to the invitation professional photographers received to cultural events that occurred in Trafalgar Square. The layered relations between image making and realising newly designed public spaces connects global competition between cities with the actions of tourists and the digital realm of online images and public spaces in London. We can conclude that rather than a demise of physical public spaces through an increase in online sites of public exchange, the online sharing of images necessitates increased and newly formed architectural public spaces to the detriment of public spaces of interaction.

Large-scale strategies

Third, I conclude that when larger-scale organisations employ strategies for comprehensive development they limit opportunities of other, often smaller scale activities, communities and actions. The cases reveal complicated relations between planners, politicians and developers, who directed the developments on metropolitan, national and international scales and individuals and small groups who operated locally within the public spaces. At Elephant and Castle, the increased confidence of the local authority partnership since 2014 undermined the potential of renovating the shopping centre, questioning the presence of the market operators and stallholders. Legal disputes between developers and leaseholders, the closure of the school for a new mixed-use development and the tension caused by unfulfilled expectations of the residents at Paddington Basin also highlight the lack of power afforded to individuals and community groups. The political significance of Trafalgar Square generated interest from national government, brand-name architects and the GLA, which I identify, impacted the presence of smaller-scale activities of vendors and performers. We can see from the three cases that the relations between organisations, people, places and planning mechanisms unfold at differing scales of space and power. While I recognise the production of public spaces in the three cases was realised across an array of intersecting scales, the power taken and leveraged by local governments (with land resources) and large-scale developers (with financial means) favoured larger scales of masterplanning and disadvantaged small-scale incremental development. Once opportunities for masterplanned redevelopment are identified, the power that councils and developers have is strongly asserted in planning and development processes. As a result, opportunities to be part of the practices of everyday and architectural transformation of public spaces become tightly prescribed.

At Elephant and Castle and Paddington Basin the limitations of the local authority and GLA planning systems were exploited by large-scale private interests. Setha Low (2006:83) states that 'private interests take over public space in countless ways'. She emphasises that in large-scale developments specific approaches are employed through physical interventions and reinforced by economic and legal strategies. Private developers recognise that local authorities are unable to realise new development projects without the presence of commercial investment. I highlighted ways in which Section 106 agreements and planning controls designed to offset the impact of large scales of development were negotiated in favour of developers and investors, often compounding the impact of the developments on residents and businesses. High-quality, new public spaces were created by developers who also made contributions towards local transport infrastructure. Such investments increased the marketability of developments for future incomers and investors while fulfilling the demands of Section 106 agreements and Community Infrastructure Levies. These contributions reduced obligations, such as the provision of social housing that offer fewer benefits to the developers.

The former owner of the shopping centre at Elephant and Castle described how their proposal to renovate the shopping centre and to add 1,000 new residential units would be offset by their financial contributions to an expanded Northern Line Tube station (Interview 2012). If implemented, the shopping centre owners would have benefitted from an improved transportation network with greater capacity and a flow of customers directly into its mall. The advantages gained by large commercial interests brings the suitability of planning control into question and raises further questions over whether there are sufficient controls over large-scale development and adequate support for public concerns and activities.

Planning negotiations conducted in private, between public agencies and governments who have land assets and large developers backed by private investors, come to define the forms and relations of masterplanned developments. In addition to governments facilitating large masterplans, private developers assert control through employing design, planning and legal consultants to negotiate their relations with politicians and council employees. What was considered by several interviewees to be the unbalanced nature of the development negotiations are not required to be shared publicly. As the significance of the financial gains of the developers increased and became more unbalanced, which were also facilitated by favourable terms agreed by the local authorities of Southwark and Westminster, it was not in the interests of either party to reveal what had been agreed. Architectural plans that were prioritised for their spaces of high aesthetic quality, as well as the long-term management of spaces, remain in the developer's control. As decisions about the future of public spaces are made in private, individuals and organisations with power use such opportunities to ensure that they benefit from what is agreed. Exclusive relations reinforce suspicions in local communities that the economic ambitions of planners and developers are prioritised over the needs and concerns of the public. As Judy Peterson, one of the former teachers from the school closed to make way for the development at Paddington Basin, described: 'the closure of the school has nothing to do with the education, it is the value of the land that is making that decision' (Interview 2013).

That unequal processes in the development of public space frequently favour large-scale, private interests was recognised by many interviewees in the research. In interviews with local authority planners, the acceptance of individuals losing out from the developments also exposed the changing roles and expectations of government. At Elephant and Castle the shopping centre owners described how they did not expect any retailers to remain after its redevelopment, particularly because the shop rents would prohibitively increase (Interview 2012). Similar comments from Tom Appleby at Southwark Council questioned the future of the market in the moat:

> I think that we would like to see opportunities for them [market traders], but whether they are the right operators to go on that new space, I don't know, it is a commercial decision in part. The one in the moat is managed by Urban Space Management, under a lease from St Modwen's, you can

> see what kind of, I'm not being derogatory to it, but if someone is going to create a new market on that space then they have to have an idea of what kind of . . . If you are setting up a new market you want to think about what ethos and the image you are creating.
>
> *(Interview 2012)*

Similarly, the actions of the security guards and the BID, who called in the police, have resulted in teenagers being evicted from the Paddington Basin area. Julian Dean of Westminster City Council explained (Interview 2013):

> While the college [North Westminster Community College] was there . . . security guards noticed people hanging around. But one of the interesting things about that is the fear of youths hanging around is much greater social factor than the actual danger of youths hanging around. And you know, with either race or dress or attitude problems, from both sides of the fence, security guards and teenagers, I couldn't imagine a worse combination, you know. But nothing that got really out of hand.

From developers, public agencies, planners and designers to tenants and residents there is an expectation, and in most cases an acceptance of the imbalances of power, that some organisations will benefit while other groups and individuals will lose out.

Future considerations

Socio-spatial relations of making

Building on the conclusions, I highlight three further considerations for realising future public spaces. The first is that, *through focusing on relations of making, social and spatial conceptions of public spaces become inseparable*. By highlighting practices of making, I have questioned narrow definitions of public spaces that serve single disciplinary approaches. I have also challenged commercial ambitions to focus public spaces on visually attractive architectural forms where social relations within them are controlled. One of the aims of the research has been to work simultaneously with architectural and social science methods in the study of public spaces. This has necessitated a consideration of public spaces as neither architectural containers within which social activities unfold nor solely public spaces formed through social interactions. What Doreen Massey terms 'thrown-togetherness' requires 'negotiation which must take place within and between both human and non-human' (2005:140). Through fieldwork I observed social relations within and as part of three specific London sites. During desk studies and interviews I further analysed the associations between the decisions and actions of using, adapting and redeveloping the physical spaces. The combination of methods revealed overlapping and competing notions and claims to public spaces, leading me to propose the adoption of a broader definition that focuses on the publicness of making public spaces undertaken by individuals and organisations with varying interests, livelihoods and disciplinary fields. This approach reflects the multiple social and spatial relations that produce public spaces.

Accepting public spaces as composed of multiple socio-spatial trajectories questions the suitability of single disciplinary tools. I propose that developers, planners, architects and landscape architects adopt research methods, such as observation, which can reveal nuanced social interactions and may lead to developments that incorporate more diverse notions of publicness. Correspondingly, I advocate that social scientists employ visual and spatial techniques while also directly proposing designs and policies for new public spaces. Such transdisciplinary approaches would bring into closer proximity reflective research methods and propositional design practices, such as the techniques of William Whyte (1980) who studied social relations within open spaces and through his conclusions sets out new coded standards for future public spaces. Whyte writes that 'If we could find out why the good plazas worked . . . we could have the basis for a new code' (Whyte 1980:15). Through openness to, and knowledge of, both social science and design methods the possibilities of research and design projects could be expanded. Such approaches could also intersect academic and professional practices and could lead to a hybridising of techniques, such as combining traditional mappings of spatial conditions undertaken by landscape architects and urban designers with conversations recorded by social scientists. The amalgamation of tools could underpin new forms of practice and representation as conversations in the field are mapped spatially and the designs of architectural spaces are proposed in writing. Most importantly, however, the combining of approaches can reveal issues, such as the making of public spaces that reveal unfair conditions experienced by small-scale activities as large-scale redevelopment unfolds. As Fran Tonkiss (2014:1) claims:

> Focusing on the interplay between the social and the physical shaping of contemporary cities makes it possible to see how the material organization of urban space is crucial to the production and reproduction of social and economic arrangements, divisions and inequalities.

There are, however, specific difficulties in combining architectural perspectives and social science approaches in the study of public spaces. While architecturally trained writers such as Madanipour (1996, 2010) and Carr et al. (1992) explore social dimensions of public spaces, when they define public space they frequently revert to notions of places developed spatially to accommodate public activities. From social science perspectives, contrasting but similarly dialogical relationships are also frequently argued for (see Low 1996, 2000 and Mitchell 1997). Through research on public plazas in Costa Rica, Low proposes a definition of space being socially produced and socially constructed (1996:861). She writes that 'the social production of space includes all those factors—social, economic, ideological, and technological—the intended goal of which is the physical creation of the material setting'. Correspondingly, she defines the social construction of public space as the 'phenomenological and symbolic experience of space as mediated by social processes such as exchange, conflict, and control' (1996:861). By working with these definitions, Low attempts to 'integrate the perspectives of social production and social construction of space' (1996:862). However, such descriptions highlight binary approaches that in the case of this research keeps sociological and architectural perspectives apart. When reflecting on these two contrasting positions in *Spatializing Culture: An Engaged Anthropological Approach to Space and Place* (2014) Low explains that the 'co-production model' that she previously employed 'was limited by its two-dimensional structure' (2014:35). Her response is to add 'embodied space' (2014:35) as a third conception, building on an interest in theories that are 'experience-near and yet allow for linkages to be made to larger, social, and cultural processes' (2003:10). In contrast, I propose a single, but broader, conception of a *publicness of making* public spaces, conceived as varying combinations of both social spaces *and* spatial lives always in process.

The potential revealed in combining social and spatial approaches highlights the need to question other commonly described assumptions of public space, such as between narrow terms of ownership (corporate and state), scale (large and small) or formation (planned and spontaneous). All three cases in the book are composed from a patchwork of spaces owned and operated differentially by private and public interests—a mix of publicly and privately owned spaces as considered in the research as public spaces, which are identified to have existed before all three developments began. However, I have also identified that large-scale mechanisms of masterplanned development have facilitated the consolidation of private ownerships at Elephant and Castle and Paddington Basin and state control has been reinforced at Trafalgar Square. When reflecting on issues of scale, by merely focusing on the impact of large-scale developments on small-scale uses, other dimensions and relations that criss-cross between them could have be missed. I revealed that mid-scale businesses, events, gatherings and operations are found to have significant presence in each of the three cases. Moreover, through unquestionably accepting the overshadowing of small activities by masterplanned developments, I could have overlooked the significant influence of some individuals and the lack of power of other larger groups.

Continually making public space

The second consideration advocates that *public space should be understood as a socio-spatial entity continually remade and unmade through differential rhythms, scales and compositions of publicness.* Considering trajectories of tourists, commuters and cleaners intermixed with seasonal markets and political events—transforming places previously reconfigured through urban design plans—reveals public spaces as 'product[s] of interrelations' which are 'always in the process of being made' (Massey 2005:11). The constant interactions between people and specific geographies form new public spaces: people gather and disperse, markets are fabricated and then dismantled, regulations open up uses of public spaces while also restricting other activities. The ways in which public spaces are transformed varies between different relations and opportunities resulting in differential degrees of publicness. Through continually making and unmaking public spaces, unequal relations between individuals, stakeholders, state agencies and commercial interests are navigated and contested. As opportunities to remake these spaces are won or lost, the publicness of these sites as public spaces is redefined once again.

If public spaces are continually made through socio-spatial relations, this increases the need to negotiate contestations in public. Mitchell states that 'public space is always a negotiation' (1997:327), however, new regulations that restrict certain activities and participants can 'remove some people from the negotiators' table' (ibid.). Everyday rhythms are interrupted by large-scale masterplans and new legislation that extinguish some activities while opening up opportunities for others. At Elephant and Castle Market the layered controls of the local authority, shopping centre owners, market managers and stallholders formed a site of cyclical daily routines and spatial configurations. Although the latest regeneration has resulted in the closure of the market in the moat, other public interactions will inevitably emerge in the new market square. Similarly, controls by private interests led by the developers at Paddington Basin narrow opportunities for public discourses and action, but have not extinguished them altogether. At Trafalgar Square daily activities continue despite the intensity of scheduled cultural, political and commercial events that occupy the square since it was refashioned in 2003. However, in contrast to the publicness within these sites, many conversations that led to agreements were undertaken in private, leaving negotiations conducted outside of public scrutiny. A lack of transparency, in the Section 106 planning agreements with developers and out-of-court settlements with individuals, has led to suspicion and accusations of injustice in all three cases. Issues that impact the formation of public spaces have the potential to be more openly negotiated in public and with more inclusive consultation.

If public spaces are accepted as being continually reconfigured and redefined, and where the relations are open to public scrutiny, there may be potential for different people to participate within the dynamics of making. Through all stages of redevelopment, daily reconfigurations of market stalls and concert stages and longer temporal phases of artworks, signage and installations were found to cut across the architectural refashioning of the spaces. The differing compositions of commuters, tourists, students, performers, market traders, protestors and construction workers intermingled with the material spaces and continually reconstituted the sites as public spaces. Occasionally, notions of the public sphere as 'arenas of social and political contest and struggle' (Smith and Low 2006:12), become indistinguishable from definitions of public space. Although this framing of public space as constantly in flux has the potential to unsettle, it also provides for opportunities to engage in diverse interactions: as places defined by the contestation of public issues, as sites to be realised through design and as spaces of community action. If sites are continually being reformed as public spaces, rather than being fixed in place through tightly controlled rhythms of management and events, then they offer the potential for many people to participate in their production in more varied and inclusive ways.

Publicness of making

Finally, I propose that *the publicness of public spaces should be considered in terms of the inclusiveness of opportunities to redefine them.* To address narrow frames of ambition for public spaces identified in the research, as *spatial forms*, *visual images* and *social interactions*, alternative definitions need to be imagined. The degree of publicness of public spaces needs to be considered in relation to the openness for different people and organisations to engage in the continual transformation of public space sites. This provocation reflects Massey's description of public spaces as places which are 'deliberately open' in which 'we have the possibility of constructing a public' (www.publicspace.org 2013). I propose that what Madanipour terms '*inclusive processes*' (2010:1, italic in original) in making public spaces, should be as important as the physical form or legal ownership in the definition of a site's publicness. We should consider public spaces in relation to their accessibility to socio-spatial processes, where people continually challenge the forms and terms through their presence, negotiations and actions. The degree to which a place could be termed a public space could depend on the openness to uses and participants, such as school children entrusted to redesign a park or tourists afforded rights to reconfigure a civic square through their engagement with buskers. Socio-spatial possibilities could result in what Low (2003, 2014) may be aiming for with her trialogue between the social production, social construction and embodied conceptions of public space: 'space as a potentiality for social relations' (2014:35).

In the context of the findings, how could future public spaces as socio-spatial entities be considered? How could open definitions of public space be achieved despite competing agendas for the spaces under constant negotiation in their production? A common question could be asked of Elephant and Castle Market, Paddington Basin and Trafalgar Square: how can an openness and equity in access to continually remake public spaces be achieved? Since concerns over privatisation of public spaces were raised in the 1980s and 1990s there has been a refocus on public space from issues of ownership to concerns for ongoing management. But despite the GLA report, *Public Life in Private Hands: Managing London's Public Space* (2011), stating that there should be attempts to 'engage the community in both the design and the ongoing management process' (2011:42), there does not appear to be a slow-down of private interests involved in making public spaces or an increase in community decision making. Commercial organisations, represented by private developers and supported by state agencies and local authorities, are increasingly well represented in the redevelopment of public spaces while local groups and individuals struggle to be involved. These are concerns explored in the following epilogue, through three propositions that embrace a diversity of notions of public spaces continually produced through more inclusive, transparent and fairer means.

EPILOGUE

Three propositions

Having worked for many years on the practice and teaching of designing large urban public landscapes, I find myself asking how different forms, practices and conceptions of public spaces can be established. In the context of this research, I also question the economic priorities of large-scale masterplanning that employs visual, spatial and regulatory approaches and I argue that we must seek out more inclusive socio-spatial practices for producing equitable urban public spaces. In these final pages, I set out three propositions that imagine how new conceptions of public space could be realised in practice. First, I propose greater transparency in how public spaces are planned, designed, built and used. Second, I advocate that a more inclusive approach to defining public spaces could be achieved through a London-wide design code for public spaces informed by individuals, community groups, government and commercial organisations. Third, I suggest that an independent regulator for London's infrastructure of public spaces could ensure transparency and inclusivity while overseeing the many claims to public spaces, ownerships, management and regulations.

Design coding is described in *Preparing Design Codes: A Practice Manual* as providing precise spatial design guidance through 'written and graphic rules' (Communities and Local Government 2006:10). David Grahame Shane explains that such urban codes are a 'written, systematic and logically consistent body of laws and statutes that regulate a situation, activity or place' (2011:348). Design coding was strongly promoted by the Commission for Architecture and the Built Environment (CABE) and the Department for Communities and Local Government, during the period of masterplanning described in this book. Government-commissioned design code documents, such as *The Use of Urban Design Codes* (CABE 2003), *Design Coding: Testing its use in England* (CABE 2005) and *Preparing Design Codes: A Practice Manual* (Communities and Local Government 2006), provide an argument for employing design coding to realise spatial and economic urban developments. They focus on the benefits that design coding offers planners, designers and developers to create: 'better designed development'; 'enhanced economic value'; 'a more certain planning process'; 'streamlined development control'; and 'more coordinated development' (2006:10). More recently *Public London: Ten Years of Transforming Space* (2015), a study of public spaces by urban development advocates New London Architecture (NLA), calls for the establishment of a 'code of practice to ensure public space is public for all' (2015:101). However, I recognise that the conditions described in the CABE and NLA documents exacerbate, rather than address, the issue that I have identified in the research—concerns that the development of public spaces is too focused on economic value addressed through spatial and visual design practices, resulting in the systematic marginalisation of certain individuals and communities. Rather than agendas only directed to benefit the interests of government officials, developers and urban designers, I propose that a new public space code could address concerns for fairness by empowering less visible, smaller scale interests and individuals.

Through the following propositions I advocate inclusive ways of developing public spaces that could be informed by diverse publics and which could further provide for, rather than deny the presence of contentious activities, undesirable individuals or unplanned gatherings. The coding of these propositions, as Michael Sorkin attests in his utopian manifesto *Local Code* (1993:127):

> embraces the idea that the city is a collaborative artefact and calls for a re-centering of the framework for such collective activity, for a re-examination of the narrow coercions of conventional "master" plans.

Design codes that are too open can be ignored or misinterpreted by developers and other stakeholders, and thus fail to fulfil the intentions for which they are authored. Alternatively, and in the context of public space in London, narrowly prescriptive codes can too closely reflect the regulations that I have found to accompany newly designed architectural public spaces. Shane recognises a history of such legal codes used in 'restricting the rights of urban minorities' (2011:349). I suggest, however, that design codes and planning policies must simultaneously open up opportunities, as Sorkin proposes (1993:127), while also achieving precision in how public spaces are produced. William Whyte, who employed a design code in his proposals for public spaces in New York City, claims that 'Lack of guidelines does not give builders and architects more freedom. It reinforces convention' (1980:30). If we were to apply Sorkin's *Local Code* to London, with a code such as, 'No public space should be unlinked. The public area of the City is to be continuous' (1993:66), we could realise a network of connected public spaces while affording landscape architects, community groups or developers the opportunity to decide how streets, squares, plazas and other spaces designated as public spaces are to be joined up. If we consider this code alongside 'Public Space may be stacked' (Sorkin 1993:66), then vertical connections to public roof gardens, such as the elevators connecting the street to London's Sky Garden at 20 Fenchurch Street, would also be required to be public spaces. Furthermore, 'Stacked Public Space may comprise no more than 10% of the total area of Public Space in the City' (Sorkin 1993:66) would provide specific limits to the development of roof gardens that are claimed to be public spaces.

While design codes can prioritise spatial forms, their impact on social relations can be considerable. The Department of Communities and Local Government considers design codes to primarily facilitate a spatial approach to redevelopment: 'Design codes support the culture change in planning and the transition to a spatial approach to planning' (2006:10). Such a spatial approach is emphasised by Sorkin: 'I have tried to include no direct prescription for the character of social relations' (1993:127). However, the aims from which design codes are written and the architectural forms that result are social as well as spatial. For example, Sorkin's code includes schools and markets (1993:52) reflecting his social priorities and indicating the potential socio-spatial relations generated from these specific places of education and exchange. Through the socio-spatial propositions, written as urban code, I aim to provide a fairer means for many different people and organisations to transparently design, regulate and use future public spaces. While it builds on the research to focus on London, its capacity for informing how public spaces are produced in cities around the world is also evident.

Proposition 1: transparent negotiations

The first proposition aims to make transparent and bring to public scrutiny the relations and agreements reached in the production of public spaces in London. Susan Fainstein recognises that 'demands for transparency, inclusion, and negotiation in public decisions' are a response to the 'inequitable impacts of urban programs resulted from blocking the voices of affected publics' (2010:24). The concealed contracts and opaque exchanges that inform masterplanning neighbourhoods and designing public spaces, particularly attempts of private interests to protect their business interests, reveal significant financial opportunities for profit-driven developers. Additionally, the details of out-of-court settlements that lead to the displacement of some people from public spaces are rarely disclosed to the public. The benefits to developers of Section 106 agreements and Compulsory Purchase Orders (CPO), the relations of BIDs with developers and the sale of public assets to commercial developers suggest that there is a lot to be gained as public spaces are produced. Correspondingly, there is a lot to lose (or potential benefits to be lost) as agreements are signed, releasing public assets to commercial interests with questionable gains for the local government and the communities they are tasked to represent. The conflicts of interests within local authorities, between responsibilities for planning control and the need to foster entrepreneurial relations with developers, leave local authorities with greater incentives to conceal their agreements than to transparently present them for public scrutiny. Therefore, the first proposition is:

1. Transcripts of *all* meetings, conversations, negotiations and agreements that inform the planning, design, development and management of public spaces should be available as publicly accessible documents.

Proposition 2: inclusive practices

The second proposition highlights the exclusion of less desirable individuals from public spaces, such as market vendors, school children, homeless people and street performers, and the potential to include a more diverse community of individuals and organisations in the making of public spaces in London. Returning to Ali Madanipour's definition of public space, as '*accessible spaces*, developed through *inclusive processes*' (2010:1, italics in original) I propose a more inclusive approach in making London's public spaces through involving a broad range of public and private interests, composed from residents, businesses, visitors, stakeholders, developers and governments. Madanipour writes: 'An inclusive process would involve a larger number of people and agencies and would spread the benefits of the process to larger parts of society' (2010:12). Despite this, he notes:

> However, if the needs and demands of the disadvantaged parts of society are not strongly represented, politically or financially, as is often the case, the process and its outcomes may not serve them at all.

I propose that the potentially simple language of design coding could be employed through an incremental and cumulative process that engages schools, charities and community groups. Such an approach could include the priorities of school children, homeless people and other marginalised groups in the making of public spaces, as well as the ambitions of developers, governments and private interests. This city-wide design code could be produced through the involvement of diverse interest groups, written in both disciplinary-specific and less professional language, directly representing previously unheard voices alongside strategies already afforded publicity. Rather than translating the aspiration of communities through traditional consultation workshops, design codes could be written and directly adopted as a means of directing the redevelopment, management and use of public spaces. While the design code would require mediation (see the following, *Proposition 3: independent regulator*), the large number of individual codes, which would constitute a potential code for public spaces, could provide many individuals and groups an opportunity to inform the process. Additionally, a design code, representing strategic ambitions for development and techniques of everyday appropriation, would always be in the process of being translated through the actions of individuals using public spaces and organisations reconfiguring them through events, architectural design and masterplans.

Regarding the inclusive aspirations of design coding, in the foreword to CABE's *Design Coding: Testing its use in England*, the deputy Prime Minister, John

Prescott, writes: 'Design codes bring together the professionals involved in planning, urban design and highways, along with local residents, landowners and developers' (2005:5). The Department of Communities and Local Government report also recognises that, 'key parties have different motivations for preparing a design code' (2006:7). However, in practice most design codes tend to be, as described by Matthew Carmona and Valentina Giordano, 'authored to facilitate planning permission for the benefit of private consultants, private developers and landowners' and they 'tend not to be openly published on local authority websites' (2013:6). I therefore propose an alternative code to communicate the desires of diverse stakeholders, including the ambitions of individuals otherwise excluded. A written code for public spaces could bring coherence to the array of private and state-initiated regulations. A design code could provide a stronger prescription than the guidance provided by *The London Plan* (GLA 2012). A design code overseen by an independent regulator, as proposed in the following proposition, could address the issue of changing roles of local, metropolitan and national governments. The inclusion of a broad range of publics in writing the design code would challenge narrow private interests of individual owners and managers and could encourage local residents, visitors and organisations to engage in the remaking of public spaces. The second proposition for public space is, therefore:

> 2. The design, development, management and use of public spaces should be informed by a public space code written by a large number and diverse set of individuals, groups and organisations (and facilitated by an independent regulator, see Proposition 3).

Proposition 3: independent regulator

The third proposition is to establish an independent regulator with the responsibility of ensuring transparency and inclusivity in the design, development, management and use of London's public space infrastructure. As stated in the first proposition, public-private agreements that lead to the remaking of public spaces could be more transparent if details of negotiations were made available to the public. However, as Fainstein explains in *The Just City* (2010), only providing transparency 'fails to confront adequately the initial discrepancy of power' which forms policies and directs decision making (2010:24). As public infrastructures have been privatised across the UK, since the 1980s, once state-owned networks, such as telecommunications and transportation, have been provided with regulators to coordinate and oversee their operations (such as Ofcom and the Office of Rail and Road). In addition to overseeing networks that are operated by different private organisations and that extend beyond single local authority areas, the regulators have set standards and have acted as ombudsmen for complaints against the organisations who privately manage resources that were previously considered public goods. To accept public space as an infrastructure could broaden the scope of public spaces to include the streets, squares, canal-sides and wider networks of publicly accessible spaces; include state and privately owned or managed public spaces; and make inseparable the social and spatial dimensions of public spaces. I propose therefore that a new independent organisation is required to regulate London's public spaces.

As I have demonstrated, rights to public spaces tend to be decided by organisations with greater access to legal, land and financial resources. A lack of regulation leaves imbalances of power in the production, management and use of public space unaddressed, allowing for an uneven control of public spaces by government and private interests. As Doreen Massey states (2005:152): 'All spaces are socially regulated in some way' and 'unregulated' public space 'leaves a heterogeneous urban population to work out for itself who really is going to have the right to be there'. In the context of the proliferation of forms and management of public spaces, an independent regulator could provide fair and consistent oversight over the operations and regulations of these spaces across London. However, rather than merely regulating the spatial forms of public spaces or the less ordered, small-scale activities of vendors, buskers, adolescents or transients, a new independent regulator could oversee all socio-spatial concerns for making public spaces, including large-scale political and commercial interests. With the fragmentation of responsibility for public space currently across the thirty-two London boroughs, within new redevelopments and historic estates, and between BIDs and private management companies, an independent regulator could provide a consistent overseeing body. An independent regulator could also ensure that public interests are fairly represented in the making of public spaces as the role of local government, leadership and priorities of the GLA and the political agendas of central government change. The third proposition is:

> 3. A new independent regulator for London public spaces should be established to represent the public (encompassing the entirety of Greater London and remaining independent from central government, the GLA, local authorities and private interests).

In concluding, I would like to reemphasise that approaches to masterplanning can create exclusive public spaces that overshadow the daily routines of sites across London. As Madanipour (2010:13) highlights:

> disadvantaged groups, who do not have access to financial resources and are frequently disconnected from the political process, end up having no control or stake in the city building process. The places that are created are not designed to serve them, as these groups are not often part of the decision-making formula.

Within large neighbourhoods and districts that are enveloped by masterplans led by governments, agencies and developers—public spaces, livelihoods and concerns of many individuals are disregarded. Behind bold designs for new urban spaces and the attraction of a reprogrammed public realm, competing interests are strongly asserted, creating unbalanced relations between commercial developers and small-scale businesses and between local authorities and activities of individuals deemed undesirable.

I argue that there is an urgency to create an independent regulator to establish transparency, to include the interests of marginalised individuals and groups and to oversee the presence of competing public and private interests in the development, management and use of London's public spaces. This research has demonstrated that even before the accelerated privatisation of public spaces in the 1980s, private interests have played a significant role in the provision and use of public spaces. To merely propose that the distinctions between public and private spaces be more clearly delineated ignores this history and overlooks the changing role of governments (national, metropolitan and local authority) and public agencies (such as the Canal and River Trust). I instead advocate for transparency of these relationships that have become increasingly important in the lives of London's public spaces.

That such relations are often reliant on large scales of redevelopment must not overshadow the diversity of individuals and activities, often smaller scale, from being represented in the making of public spaces. I would like to reiterate that while public space is an essential resource that is repeatedly claimed and transformed by a range of powerful public and private interests, there are potentially fairer ways of opening access to these processes through independent regulation. I do not merely propose a further addition of regulations and laws, but rather a structure that opens up new opportunities for inclusively planning, managing and using public spaces. This could be achieved, in part, by a design code that realises physical public spaces as inclusive processes. Such an approach, of bringing together many short sentences of design code to create a collective strategy for public space, as Sorkin states in *Local Code* (1993:11), could recognise that: 'The City will always prefer to see the small initiative reflected in the large'.

BIBLIOGRAPHY

35% Campaign. *Southwark Council's Revolving Doors.* www.35percent.org/revolving-doors/ (accessed 4 September 2016).

Akkar Ercan, M. 2010. Less public than before? Public space improvement in Newcastle City Centre. In: A. Madanipour (ed.) *Whose public space?* Oxon: Routledge.

Althrope, M. 2008. The elephant in context. *The Car and the Elephant.* www.thecarandtheelephant.com/chapter/context (accessed 12 December 2012).

Amin, A. 2008. Collective culture and urban public space. *City: Analysis of Urban Trends, Culture, Theory, Policy, Action*, 12 (1), 5–24.

Atkinson, R. and Flint, J. 2001. Accessing hidden and hard-to-reach populations: Snowball research strategies. *Social Research Update*, 33.

Augé, M. 1995. *Non-places: Introduction to an anthropology of supermodernity.* London: Verso.

Barthes, R. 1981. *Camera Lucida: Reflections on photography.* New York: Farrar, Straus and Giroux.

Bates, C. 2009. The Eiffel Tower is the world's most photographed landmark . . . but London's Trafalgar Square comes a close second. *Mail Online.* www.dailymail.co.uk/sciencetech/article-1175055/The-Eiffel-Tower-worlds-photographed-landmark-Londons-Trafalgar-Square-comes-close-second.html (accessed 12 December 2012).

Becker, H.S. 1998. *Tricks of the trade: How to think about your research while you're doing it.* Chicago, IL: University of Chicago Press.

Becker, H.S. and Faulkner, R. 2008. Studying something you are part of. *Ethnologie Francaise*, 38, 15–21.

Berger, J. 2013. *Understanding a photograph.* London: Penguin Classics.

Boyer, C. 1992. Cities for sale: Merchandising history at south street seaport. In: M. Sorkin (ed.) *Variations on a theme park.* New York: Noonday Press.

British Sociological Association. 2002. Statement of ethical practice. *Broadgate Estates, Public Realm Estate Management.* www.broadgateestates.co.uk/public-realm-estate-management (accessed 2 May 2013).

CABE. 2003. *The use of urban design codes: Building sustainable communities.* London: Commission for Architecture and the Built Environment.

CABE. 2005. *Design coding: Testing its use in England*. London: Commission for Architecture and the Built Environment.

CABE Space. 2004. *The manifesto for better public spaces*. London: Commission for Architecture and the Built Environment

Campkin, B. 2013. *Remaking London*. London: IB Taurus.

Canal and River Trust: Planning and Design. https://canalrivertrust.org.uk/about-us/planning-and-design (accessed 1 August 2016).

Carmona, M. 2003. *Public places, urban spaces: The dimensions of urban design*. Oxford: Architectural Press.

Carmona, M., de Magalhães, C. and Hammond, L. (eds) 2008. *Public space: The management dimension*. Oxon: Routledge.

Carmona, M. and Giordano, V. 2013. *Design coding, diffusion of practice in England*. London: UCL and Urban Design Group.

Carmona, M. and Wunderlich, F.M. 2012. *Capital spaces: The multiple complex public spaces of a global city*. Oxon: Routledge.

Carr, S.M., Francis, M., Rivlin, L.G., et al. 1992. *Public space*. Cambridge: Cambridge University Press.

Colley, F. 2013. Elephant and castle regeneration will transform the area not just the estate. *The Guardian*. www.theguardian.com/local-government-network/2013/sep/23/elephant-castle-regeneration-heygate-estate (accessed 19 January 2014).

Communities and Local Government. 2006. *Preparing design codes: A practice manual*. London: RIBA Publishing

Corner, J. (ed.) 1999. *Recovering landscape: Essays in contemporary landscape architecture*. New York: Princeton Architectural Press.

Crandall, D., Backstrom, L., Huttenlocher, D., et al. 2009. *Mapping the worlds photos*. www.cs.cornell.edu/~dph/papers/photomap-www09.pdf

Cresswell, T. 2004. *Place: A short introduction*. Oxford: Blackwell.

Creswell, J.W. 2003. *Research design: Qualitative, quantitative, and mixed methods approaches*. Thousand Oaks, CA: SAGE.

Davies, M.J. 2007. *Property: Meanings, histories and theories*. Oxon: Routledge.

Davis, M.J. 1990. *City of quartz*. London: Verso.

De Certeau, M. 1984. *The practice of everyday life*. Berkeley, CA: University of California Press.

Degan, M. 2008. *Sensing cities: Regenerating public life in Barcelona and Manchester*. Oxon: Routledge.

Denscombe, M. 2010. *The good research guide: For small-scale social research projects*. London: The Open University Press.

Department for Communities and Local Government. 2006. *Planning obligations: Practical Guidance*. www.gov.uk/government/uploads/system/uploads/attachment_data/file/7770/151363.pdf (accessed 19 December 2013).

Department of Communities and Local Government. 2012. *Renegotiation of section 106 planning obligations: Consultation*. www.gov.uk/government/uploads/system/uploads/attachment_data/file/8338/2196058.pdf (accessed 5 September 2015).

DETR and CABE. 2000. *By design: Urban design in the planning system: Towards better practice*. London: CABE.

Duneier, M. 1999. *Sidewalk*. New York: Farrar, Straus and Giroux.

Dyckoff, T. 2012. Let's move to Elephant & Castle and Walworth, South London. *The Guardian*. www.guardian.co.uk/money/2012/jul/06/lets-move-to-elephant-and-castle-walworth# (accessed 4 September 2016).

Elephant and Castle Traders' Association. 2007. *Elephant and Castle traders' charter* (Unpublished).

Enoch, N. 2013. Nelson would turn in his grave: Campaign to stop giant French blue cockerel from landing on Trafalgar Square's fourth plinth. *Mail Online*. www.dailymail.co.uk/news/article-2317574/Fourth-plinth-Giant-French-cockerel-set-land-unless-campaigners-stop-it.html (accessed 4 September 2016).

Fainstein, S. 1994. *City builders: Property, politics, and planning in New York and London*. Oxford: Blackwell.

Fainstein, S. 2010. *The just city*. London: Cornell University Press.

Foster and Partners. 2003. *Transformation of Trafalgar Square*. www.fosterandpartners.com/news/archive/2003/07/transformation-of-trafalgar-square/ (accessed 4 September 2016).

Fraser, N. 1990. Rethinking the public sphere: A contribution to the critique of actually existing democracy. *Social Text*, 25/26, 56–80.

Gardiner, J. 2008. *Lend Lease May Need Bail-Out Funding for Olympic Village 2008*. www.building.co.uk/news/lend-lease-may-need-bail-out-funding-for-olympic-village/3119078.article (accessed 4 September 2016).

Gehl, J. 1987. *Life between buildings: Using public space*. New York: Van Nostrand Reinhold.

Gehl, J. 2010. *Cities for People*. Washington: Island Press.

Gehl, J. and Svarre, B. 2013. *How to study public spaces*. Washington: Island Press.

Gensler and ULI. 2011. *Open space: An asset without a champion*. https://www.gensler.com/doc/survey-open-space (accessed 5 February 2022).

Gospodini, A. 2002. European cities in competition and the new 'uses' of urban design. *Journal of Urban Design*, 7 (1), 59–73.

Graham, S. and Marvin, S. 2001. *Splintering urbanism: Networked infrastructures, technological mobilities and the urban condition*. London: Routledge.

Greater London Authority (GLA). 1999. *Greater London Authority Act 1999*. https://www.legislation.gov.uk/ukpga/1999/29/contents (accessed 5 February 2022).

Greater London Authority (GLA). 2004. *London Plan 2004*. www.london.gov.uk/what-we-do/planning/london-plan/past-versions-and-alterations-london-plan (accessed 4 September 2016).

Greater London Authority (GLA). 2009. *A Manifesto for Public Space: London's Great Outdoors*. https://www.london.gov.uk/sites/default/files/londons_great_outdoors_-_mayors_manifesto.pdf (accessed 31 January 2022).

Greater London Authority (GLA). 2011. *Public Life in Private Hands: Managing London's Public Spaces*. https://www.london.gov.uk/sites/default/files/gla_migrate_files_destination/Public%20space%20June%202011%20Webme.pdf (accessed 31 January 2022).

Greater London Authority (GLA). 2012. *London Plan 2012*. www.london.gov.uk/what-we-do/planning/london-plan/current-london-plan (accessed 4 September 2016).

Greater London Authority (GLA). *Fourth Plinth*. www.london.gov.uk/what-we-do/arts-and-culture/art-and-design/fourth-plinth (accessed 4 September 2016).

Greater London Authority (GLA) (London Assembly Planning and Housing Committee). 2011a. *Combined Feedback Received, on Public Space in Private Hands—Managing London's Public Space*. https://www.london.gov.uk/sites/default/files/gla_migrate_files_destination/Public%20Space%20Report%20Feedback%20Combined.pdf (accessed 5 February 2022).

Greater London Authority (GLA) (London Assembly Planning and Housing Committee). 2011b. *Public Life in Private Hands—Managing London's Public Space*. https://www.london.gov.uk/sites/default/files/gla_migrate_files_destination/Public%20space%20June%202011%20Webme.pdf (accessed 5 February 2022).

Greater London Authority (GLA). *Trafalgar Square*. www.london.gov.uk/priorities/arts-culture/trafalgar-square (accessed 2 May 2013).
Greater London Authority (GLA). *World Squares for All*. www.london.gov.uk (accessed 2 July 2014).
Habermas, J. 1962 (translated 1989). *The structural transformation of the public sphere: An inquiry into a category of bourgeois society*. Cambridge: Polity Press.
Hall, S. 2012. *City, street and citizen: The measure of the ordinary*. Abington: Routledge.
Harvey, D. 1989. From managerialism to entrepreneurialism: The transformation in urban governance in late capitalism. *Geografiska Annaler: Series B, Human Geography*, 71 (1), 3–17.
Harvey, D. 1996. *Justice, nature and the geography of difference*. Oxford: Blackwell.
Harvey, D. 2005. *A brief history of neoliberalism*. Oxford: Oxford University Press.
Hodgson, N. 2013. Giant blue French cockerel set to be installed on Trafalgar Square's fourth plinth. *Evening Standard*. www.standard.co.uk/news/london/giant-blue-french-cockerel-set-to-be-installed-on-trafalgar-squares-fourth-plinth-8596701.html (accessed 16 May 2013).
Imrie, R., Lees, L. and Raco, M. (eds) 2009. *Regenerating London: Governance, sustainability and community in a global city*. London: Taylor and Francis.
Imrie, R. and Raco, M. 2003. Community and the changing nature of urban policy. In: R. Imrie and M. Raco (eds) *Urban renaissance?* Cambridge: Polity Press.
Jackson, J.B. 1984. The word itself. In: J.B. Jackson (ed.) *Discovering the vernacular landscape*. New Haven, CT: Yale University Press.
Jacobs, J. 1961. *The death and life of great American cities*. New York: Random House.
Kayden, J. 2000. *Privately owned public spaces: The New York experience*. New York: Wiley.
Kelso, P. 2000. Mayor attacks generals in the battle for Trafalgar Square. *The Guardian*. www.theguardian.com/uk/2000/oct/20/london.politicalnews (accessed 4 September 2016).
Knowles, C. 2000. *Bedlam on the streets*. London: Routledge.
Lees, L. 2013. Regeneration in London has pushed out poor families. *The Guardian*. www.theguardian.com/local-government-network/2013/aug/29/mixed-communities-plan-government-regeneration (Accessed 19 January 2014).
Lefebvre, H. 1991. *The production of space*. Oxford: Basil Blackwell.
Lend Lease. 2013. *Financial Report 2013*. http://phx.corporate-ir.net/External.File?item=UGFyZW50SUQ9MjA0NTA3fENoaWxkSUQ9LTF8VHlwZT0z&t=1 (accessed 19 September 2015).
Lend Lease. *Background to the Regeneration*. www.elephantandcastle.org.uk/pages/regeneration_change/130/background_to_the_regeneration_programme.html (accessed 5 January 2013).
Lend Lease. *Elephant and Castle Regeneration*. www.elephantandcastle.org.uk (accessed 1 September 2014).
London SE1: Community Website. 2007. *Lend Lease Chosen as Elephant and Castle Development Partner*. www.london-se1.co.uk/news/view/2835 (accessed 3 November 2017).
Londoners get new shopping centre at Elephant and Castle. 1965 (3 April). *Illustrated London News*. http://www.arthurlloyd.co.uk/Elephant/ElephantAndCastleNewShoppingCentre1965.htm (Accessed 8 March 2022).
Long, K. 2010. Paddington laid bare. *London Evening Standard*. www.standard.co.uk/lifestyle/paddington-laid-bare-6542246.html?dxwIndex=1 (accessed September 2016).
Low, S. 1996. Spatializing culture: The social production and the social construction of public space in Costa Rica. *American Ethnologist*, 23 (4), 861–879.

Low, S. 2000. *On the plaza: The politics of public space and culture.* Austin, TX: University of Texas Press.

Low, S. 2003. Embodied space(s): Anthropological theories of body, space, and culture. *Space and Culture*, 6 (9).

Low, S. 2006. How private interests take over public space: Zoning, taxes, and incorporation of gated communities. In: S. Low and N. Smith (eds) *The politics of public space.* London: Routledge.

Low, S. 2014. Spatializing culture: An engaged anthropological approach to space and place. In: J.J. Gieseking, et al. (eds) *The people, place and space reader.* New York: Routledge.

Low, S. and Smith, N. (eds) 2006. *The politics of public space.* New York: Routledge.

Lynch, K. 1960. *Image of the city.* London: MIT Press.

Lynch, K. 1981. *A theory of good city form.* Cambridge, MA: MIT Press.

Lynch, K. and Rodwin, L. 1958. A theory of urban form. *Journal of the American Institute of Planners*, 24 (4), 201–214.

Mace, R. 2005. *Trafalgar Square: Emblem of empire.* Cambridge: Cambridge University Press.

Madanipour, A. (ed.) 2010. *Whose public space?* Oxon: Routledge.

Madanipour, A. 1996. *Design of urban space: An inquiry into socio-spatial process.* Chichester: Wiley.

Madden, D. 2010. Revisiting the end of public space: Assembling the public in an urban park. *City and Community*, 9 (2), 265–284.

Massey, D. 1999. Negotiating disciplinary boundaries. *Current Sociology*, 47 (2), 5–12.

Massey, D. 2005. *For space.* London: SAGE.

Massey, D. 2007. *World city.* Cambridge: Polity Press.

Massey, D. 2013. Space and power are intimately related: Interview with Doreen Massey. *Public Space.* www.publicspace.org/en/post/space-and-power-are-intimately-related (accessed 4 September 2016).

Mayor of London (Greater London Authority). 2009. *A Manifesto for Public Space: London's Great Outdoors.* https://www.london.gov.uk/sites/default/files/londons_great_outdoors_-_mayors_manifesto.pdf (accessed 31 January 2022).

McGhie, C. 2001. The new arrivals at Paddington. *The Telegraph.* www.telegraph.co.uk/finance/property/4814516/The-new-arrivals-at-Paddington.html (accessed 4 September 2016).

Merchant Square. Paddington. www.merchantsquare.co.uk

Merrifield, A. 1996. Public space: Integration and exclusion in urban life. *City: Analysis of Urban Trends, Culture, Theory, Policy, Action*, 1 (5–6), 57–72.

Minton, A. 2006. *What kind of world are we building? The privatisation of public space.* London: RICS.

Minton, A. 2012. *Ground control: Fear and happiness in the twenty-first century city.* London: Penguin.

Mitchell, D. 1995. The end of public space? People's park, definitions of the public, and democracy. *Annals of the Association of American Geographers*, 85 (91), 108–133.

Mitchell, D. 1997. The annihilation of space by law: The roots and implications of anti-homeless laws in the United States. *Antipode*, 29 (3), 303–335.

Mitchell, D. 2003. *Right to the city: Social justice and the fight for public space.* New York: The Guilford Press.

Mitchell, D. and Staeheli, L. 2006. Clean and safe? Property redevelopment, public space, and homelessness in Downtown San Diego. In: S. Low and N. Smith (eds) *The politics of public space.* New York: Routledge.

Mitchell, D. and Staeheli, L. 2008. *The people's property? Power, politics and the public*. Oxon: Routledge.

Moore, R. 2009. Broadgate tower. *The Architect's Journal*. www.architectsjournal.co.uk/office/broadgate-tower-city-of-london-by-som/1995622.article (accessed September 2016).

New London Architecture (NLA). 2012. *Investing in Southwark: On location conference*. 19 July 2012. London: Unicorn Theatre.

New London Architecture (NLA). 2015. Public London: Ten years of transforming space North Westminster Community School, Year 13.2008. *The Beginning of the End*. www.youtube.com/watch?v=i65M4dxbxkc (accessed 16 May 2013).

Paddington Now (BID). www.paddingtonnow.co.uk/about/ (accessed 5 September 2015).

Paddington Regeneration Partnership (PRP). 2001. Paddington Waterside: Creating a place Paddington Waterside Partnership. *Paddington Waterside*. www.paddingtonwaterside.co.uk (accessed 23 October 2013).

Philby, C. 2010. Living in Ghostland—the last Heygate residents. *The Independent*. www.independent.co.uk/news/uk/home-news/living-in-ghostland-the-last-heygate-residents-1930054.html (accessed 19 January 2014).

Planning and Compulsory Purchase Act 2004 [Online]. London. https://www.legislation.gov.uk/ukpga/2004/5/contents (accessed 3 February 2022)

Poynter, G. 2009. The 2012 Olympic Games and the reshaping of East London. In: R. Imrie, L. Lees and M. Raco (eds) *Regenerating London: Governance, sustainability and community in a global city*. London: Taylor and Francis.

Raco, M. and Henderson, S. 2009. Flagship regeneration in a global city: The re-making of Paddington Basin. *Urban Policy and Research*, 27 (3), 301–314.

Ragin, C.C. 1992. Introduction: Cases of "What is a case?" In: C.C. Ragin and H.S. Becker (eds) *What is a case: Exploring the foundations of social enquiry*. New York: Cambridge University Press.

Ragin, C.C. and Becker, H.S. 1992. *What is a case: Exploring the foundations of social enquiry*. New York: Cambridge University Press.

Roche, M. 2002. *Mega-events and modernity: Olympics and expos in the growth of global culture*. London: Routledge.

Rogers, R. and Fisher, M. 1992. *A new London*. London: Penguin Books.

Rowe, C. and Koetter, F. 1978. *Collage city*. Cambridge, MA: MIT Press.

Searle, A. 2012. Off their rockers: The artists behind the latest addition to the fourth plinth. *The Guardian*. www.theguardian.com/artanddesign/2012/feb/19/elmgreen-dragset-trafalgar-fourth-plinth (accessed 4 September 2016).

Sennett, R. 1977. *The fall of public man*. New York: Knopf.

Sennett, R. 1994. *Flesh and stone: The body in western civilization*. London: Faber and Faber.

Shane, D.G. 1995. Balkanization and the postmodern city. In: P. Lang (ed.) *The mortal city*, 54–69. New York: Storefront Books.

Shane, D.G. 2005. *Recombinant urbanism: Conceptual modelling in architecture and city theory*. Chichester: Wiley.

Shane, D.G. 2011. *Urban design since 1945*. Chichester: Wiley.

Smith, N. 1992. 'New city, new frontier'. In: M. Sorkin (ed.) *Variations on a theme park: The new American city and the end of public space*, 61–93. New York: Hill and Wang.

Smith, N. 1996. *The new urban frontier: Gentrification and the revanchist city*. London: Routledge.

Smith, N. and Low, S. 2006. Introduction: The imperative of public space. In: S. Low and N. Smith (eds) *The politics of public space*. New York: Routledge.

Soja, E. 1995. *Thirdspace*. Oxford: Blackwell.

Sontag, S. 1977. *On photography*. London: Penguin.

Sorkin, M. (ed.) 1992. *Variations on a theme park: The New American City and the end of public space*. New York: Hill and Wang.

Sorkin, M. 1993. *Local code*. New York: Princeton Architectural Press.

Sorkin, M. 2009. *Twenty two minutes in Manhattan*. London: Reaktion Books.

Sorkin, M. 2011. The end(s) of urban design. In: M. Sorkin (ed.) *All over the map: Writing on buildings and cities*. New York: Verso.

Southwark Council. 2012. The Elephant and Castle opportunity area supplementary planning document (SPD). London: Southwark Council.

Southwark Council. 2013. *Council will insist on demolition*. www.southwark.gov.uk/news/article/1368/council_will_insist_on_demolition_of_shopping_centre_at_elephant (accessed 19 January 2014).

Southwark Council. *Elephant Park: Former Heygate residents 'Right to Return'*. www.southwark.gov.uk/info/200183/elephant_and_castle/1124/elephant_park/4 (accessed 4 September 2016).

Southwark Council. *Section 106 annual report 2012–14*. www.southwark.gov.uk/download/downloads/id/13876/s106_annual_report_2012-14 (accessed 5 February 2017).

Southwark Notes. 2013. Council leaks Southwark/Lend Lease Confidential Regeneration Agreement. *Southwark Notes*. https://southwarknotes.wordpress.com/2013/02/04/council-leaks-southwark-lend-lease-confidential-regeneration-agreement/ (accessed 19 September 2015).

Southwark Notes. Heygate estate. *Southwark Notes*. http://southwarknotes.wordpress.com/heygate-estate/ (accessed 13 January 2013).

St. Modwen. 2013. *St. Modwen joint venture sells Elephant & Castle Shopping Centre for £80m to DV4 and APG*. www.stmodwen.co.uk/media-centre/news/view/st-modwen-joint-venture-sells-elephant--castle-shopping-centre-for-80m-to-dv4-and-apg (accessed 4 September 2016).

Staeheli, L. and Mitchell, D. 2008. *The peoples property? Power politics and the public*. New York: Routledge.

Steadman, I. 2013. Look to the Heygate Estate with what is wrong with London's housing. *New Statesman*. www.newstatesman.com/politics/2013/11/look-heygate-estate-whats-wrong-londons-housing (accessed 19 September 2015).

Sudjic, D. 2003. So Nelson, what's your view? *The Guardian*. www.theguardian.com/theobserver/2003/jun/29/2 (accessed 4 September 2016).

The Artworks. 2017. *Elephant social market*. www.theartworks.london/elephant-social/ (accessed 7 November 2017).

The Business Improvement District (England) Regulations 2004. [Online]. London. https://www.legislation.gov.uk/uksi/2004/2443/contents/made (accessed 3 February 2022)

The Ten Most Photographed Places on Earth. 2009. http://travel.yahoo.com/p-interests-40218938 (accessed 2 May 2013).

The Paddington Partnership. 2017. *This is Paddington, the story of Paddington: A place for people*. www.thisispaddington.com/the-story-of-paddington (accessed 7 November 2017).

Tonkiss, F. 2005. *Space, the city and social theory: Social relations and urban forms*. Cambridge: Polity Press.

Tonkiss, F. 2014. *Cities by design: The social life of urban form*. Cambridge: Polity Press.

Town and Country Planning Act. 1990. www.legislation.gov.uk/ukpga/1990/8/section/106 (accessed 5 September 2015).

Transport for London. 2007. *Legible London: Yellow book*. London: Applied Information Group.

Transport for London. 2009. *Transport for London's guardrail removal scheme*. https://content.tfl.gov.uk/Item09-Guardrail-Removal-Programme.pdf (accessed 5 February 2022).

Urban 75. 2013. Elephant and Castle Regeneration Outrage. *Urban 75*. www.urban75.org/blog/elephant-and-castle-regeneration-outrage-2535-new-homes-and-just-79-social-rented-units/ (accessed 19 January 2014).

Urban Space Management. www.urbanspace.com (accessed 4 September 2016).

Urban Task Force. 1999. *Towards an urban renaissance*. London: Spon Press.

Urban Task Force. 2005. *Towards a strong urban renaissance*. London: Urban Task Force.

Urry, J. 2001. Globalising the tourist gaze. In: *Cityscapes Conference*, Graz, November 2001. Published by the Department of Sociology, Lancaster University. www.comp.lancs.ac.uk/sociology/papers/Urry-Globalising-the-Tourist-Gaze.pdf

Urry, J. and Larsen, J. 2011. *The tourist gaze 3.0*. London: SAGE.

Visit London. *Top 10 film locations in London*. www.visitlondon.com/things-to-do/sightseeing/film-locations/top-10-film-locations-in-london (accessed 4 September 2016).

Waterman, T., Wolff, J. and Wall, E. 2021. *Landscape citizenships*. Oxon: Routledge.

Watson, S. 2006. *City publics: The (dis)enchantments of urban encounters*. Oxon: Routledge.

Watson, S. and Studdert, D. 2006. *Markets as sites for social interaction: Spaces of diversity*. Bristol: Policy Press.

Whitehead, J. 2012. A walk around the modern Paddington Basin in 2012. *Local Local History*. www.locallocalhistory.co.uk/mp/walk2/index.htm (accessed September 2016).

Whyte, W.H. 1980. *The social life of small urban spaces*. New York: Project for Public Spaces.

Whyte, W.H. 1988. *City: Rediscovering the center*. New York: Doubleday.

Worpole and Knox. 2007. *The social value of public spaces*. York: The Joseph Rowntree Foundation.

Zukin, S. 1991. *Landscapes of power: From Detroit to Disney World*. Berkeley, CA: University of California Press.

Zukin, S. 1995. *The cultures of cities*. Cambridge, MA: Blackwell.

Zukin, S. 2009. *Naked city: The death and life of authentic urban places*. New York: Oxford University Press.

APPENDIX A

List of interviews and conversations[1]

In-depth interviews

Paul, resident and employee at London Borough of Southwark. 5 March 2012
Simon Gregory, Elephant and Castle resident and historian. 9 March 2012
Tom Appleby, planner at London Borough of Southwark. 23 March 2012
Jonathan Newton, Elephant and Castle Shopping Centre manager. 30 March 2012
Jane Goodwin, Chair of Elephant and Castle community group. 30 March 2012
David, architect involved in design of Elephant and Castle Shopping Centre. 2 May 2012
Jason, planning consultant. 15 May 2012
Steven James, Elephant and Castle manager. 25 May 2012
Roland, Elephant and Castle market manager. 25 May 2012
Edward Shoemaker, Paddington Basin resident. 10 October 2012
Julian Dean, planner at City of Westminster. 22 October 2012
Sharon Peters, Paddington Basin resident. 24 October 2012
Kate Beaton, Paddington Partnership and Paddington Business Improvement District. 3 December 2012
Michael Bond, Canal and River Trust. 11 January 2013
Peter Moorland, architect. 16 January 2013
Judy Peterson, former schoolteacher. 5 April 2013
Christy McLean, Greater London Authority. 10 October 2013
Sam, architect. 7 November 2013
Jason Cobb, union organiser. 15 November 2013
Michael Bond, photographer. 15 July 2013
Jason, event planner. 22 October 2013
Simon Townsend, architect. 2 April 2014
John Prescott, Deputy Prime Minister. 1 November 2014

Conversations

Barry, Elephant and Castle market trader. 25 May 2012
Al, Elephant and Castle market trader. 25 May 2012
Gregor, Elephant and Castle market trader. 7 August 2012
[unknown], Elephant and Castle security guard. 8 September 2012
Simon, Paddington Basin security guard. 6 October 2012
[unknown], Paddington Basin security guard. 6 October 2012
Steve, Elephant and Castle market trader. 14 January 2013
Ahmed, Elephant and Castle market trader. 14 January 2013
[unknown], Paddington Basin security guard. 15 March 2013
Humphrey, Paddington Basin market manager. 4 April 2013
[unknown], Paddington resident fishing on canal. 4 April 2013
[unknown], tourist. 4 April 2013
Maria, Paddington Basin market stall holder. 11 April 2013
[unknown], falconer. 6 June 2013
[unknown], security guard. 28 June 2013
[unknown], tourist. 3 July 2013
[unknown], cleaner. 3 July 2013
[unknown], security guard. 3 July 2013
[unknown], Heritage Warden. 4 July 2013
[unknown], photographer. 25 July 2013
[unknown], reporter. 25 July 2013
[unknown], tourist. 8 August 2013
[unknown], language teacher. 21 July 2015

Note

1. All names of interviewees throughout the book have been anonymised, except for that of Deputy Prime Minister John Prescott who during the re-planning of Trafalgar Square was Secretary of State for the Environment, Transport and the Regions and responsible for planning. The high-profile role that Deputy Prime Minister John Prescott had during this period makes it impractical to anonymise his account.

INDEX

Note: Page numbers in *italic* indicate a figure on the corresponding page.